THE UP-TO-DATE
HARDWOOD FINISHER

IN TWO PARTS
ILLUSTRATED

PART ONE

GIVING RULES AND METHODS FOR WORKING HARD-
WOODS, WITH DESCRIPTION OF TOOLS REQUIRED, THE
METHODS OF USING, AND HOW TO SHARPEN AND CARE
FOR THEM, INCLUDING SAWS, PLANES, FILES, SCRAPERS,
CHISELS, GOUGES AND OTHER WOOD-WORKING TOOLS
HOW TO CHOOSE HARDWOODS FOR VARIOUS PURPOSES,
AND HOW TO WORK AND PROPERLY MANAGE VENEERS
THE PROPER USE OF GLUE, DIRECTIONS FOR PREPAR-
ING GLUE, BLIND OR SECRET NAILING, HOW DONE AND
HOW FINISHED   HOW TO SHARPEN AND USE SCRAPERS
OF VARIOUS FORMS, WITH ILLUSTRATIONS SHOWING
TOOLS AND HOW TO HANDLE THEM PROPERLY, ETC

PART TWO

TREATS ON THE FILLING, STAINING, VARNISHING, POL-
ISHING, GILDING, ENAMELING AND FINISHING ALL
KINDS OF WOODWORK   IT ALSO TREATS ON RENOVAT-
ING OLD WORK, RE-POLISHING, RE-VARNISHING AND
WOOD FINISHING GENERALLY   THERE IS A SHORT
TREATISE ON DYEING WOODS IN VARIOUS COLORS FOR
INLAYING AND MARQUETRY WORK, WITH RULES FOR
MAKING STAINING, DYES, FILLERS, AND POLISHES OF
VARIOUS KINDS, FRENCH POLISHING, HARD-OIL FINISH.
RUBBED AND FLAT FINISH, TREATMENT OF HARD-
WOOD FLOORS, WAXING, POLISHING, SHELLACKING
AND GENERAL FINISHING OF HARDWOOD IN ALL CON-
DITIONS

By FRED T. HODGSON, ARCHITECT

*Member of Ontario Association of Architects, Editor of "National
Builder," and author of the "Modern Estimator and
Contractors' Guide" "Modern Carpentry," "Archi-
tectural Drawing Self-Taught," etc.*

Fredonia Books
Amsterdam, The Netherlands

The Up-To-Date Hardwood Finisher

by
Fred T. Hodgson

ISBN: 1-4101-0155-X

Reprinted from the 1915 edition

Fredonia Books
Amsterdam, The Netherlands
http://www.fredoniabooks.com

In order to make original editions of historical works
available to scholars at an economical price, this
facsimile of the original edition of 1915 is
reproduced from the best available copy and has
been digitally enhanced to improve legibility, but the
text remains unaltered to retain historical
authenticity.

The Up-to-Date Hardwood Finisher

INTRODUCTORY

Wood is, and always has been, one of the most important and useful materials that Dame Nature has vouchsafed to bless humanity with, and the latter's necessities and ingenuity have made the best of the circumstances.

It need hardly be said that Nature seemed to have well understood the wants of her children and provided for them in a most liberal manner, for it is said that nearly one-third of the earth's surface is covered by trees; all of which are, in some form or another, contributory to the wants and pleasures of man.

The introduction of machinery for the rapid working of nearly all kinds of intricate mouldings and shapes of woodwork has, in a great measure, had a tendency to elevate the taste of the whole community, though I must confess that artistic excellence of a high order, in the mechanical arts, is now scarcely found among our younger workmen, machinery having almost done away with the necessity for the fine kinds of hand and brain work. Fashion, which rules despotically in the wardrobe, influences, to a greater or lesser degree, the style and finish of woodwork and, to a certain extent, the kind of wood that must be used for certain purposes. Thirty or forty years ago no other

7

wood than clean white pine would be permitted to do service in a building. All woodwork had to be pine; floors, doors, windows, even to the bath fittings. Then Dame Fashion sent forth her decree and a mixture of white and Southern pine was used, followed shortly afterward by the abominable mixture of ash, walnut and chestnut. Some architect, who deserves well of his country, introduced all walnut and all mahogany fitments; and at once people of taste who saw this manner of finish noticed its superiority over the medley; and the fashion then of finishing in one kind of wood became the rage. Walnut was found too dark for general purposes and was soon abandoned for the lighter woods; chestnut, sycamore, ash, cherry, birch and oak were then tried, but I believe I voice the opinion of a majority of architects when I say that, with the exception of oak, the light-colored woods were not successful, from an æsthetic point of view, and it is on record that many buildings finished in these woods have since had their woodwork cleaned and stained to imitate darker woods or have been painted. One of the valuable qualities of white oak is, that it grows richer in color as it gets older, and no matter how it is finished, so long as the grain is visible, it mellows and improves with age. This is a quality that no other of our American woods possesses in the same degree. All the oaks have this quality to a certain extent, but the white oak (*Quercus alba*) possesses it more than any other. Mahogany, too, has this quality largely, if true Spanish mahogany is used, but little of this is in the market; though there are many woods that have the appearance of mahogany, and are called mahogany, such as cherry and black birch, and both buyer and user are oftentimes deceived, and pay for what they do not get.

Americans have often been reproached with willfully or ignorantly ignoring or destroying their own handsome woods, while importing from abroad at excessive cost, and the payment of heavy duties, foreign woods which are much inferior to many of native growth. This folly prevailed at a time when it was fashionable and even popular to believe there was no merit in domestic productions of any sort. This state of things, however, is now at an end, and in the neighborhood of all large towns, and in country places as well, a finish in hardwood is the rule, and a finish in pine the exception, if the building to be finished makes any pretension of being "up to the times."

With regard to the difference in cost between a finish in the best clear pine, and the best selected hardwood, there is really but little, if any, if we take into consideration durability and good taste. As between pine wood and good, well finished hardwood, the disparity in value and merit is so very little as to completely silence any comparison. Between poor pine and poor hardwood the preference should be by all means given to the former; because poor pine as it reveals its defects can be puttied and painted in a manner to disguise them; whereas the defects of poor hardwood are almost incurable. The rationale of the subject seems to resolve itself into the following statement:

Modern taste in expensive dwellings calls for the free use of hardwoods. It is immaterial which are used, but highly essential that the best seasoned woods should be selected; and, further, that they should be skillfully treated and finished.

The principal recommendation of hardwood is that it admits of a treatment which renders it impervious to the effects of atmospheric changes and, therefore, can be made more durable and ultimately less expensive than pine wood. A hardwood that is well seasoned before use, that is treated with proper fillers to close up its pores, and then finished with successive coats of suitable varnish, well rubbed in with pumice-stone, being finally brought to a higher flat finish, presents the most attractive, serviceable and reliable style of woodwork than can be introduced into a house. Well finished hardwood obviates the expense and annoyance of constant renewals, which pine work calls for in patching, puttying and painting. The most brilliant gloss of finished hardwood can be restored by a skilled person without disturbing the furniture or carpets of a house.

Pine work seems peculiarly and incomparably adapted for cheap work. A good article of common pine, suitable for ordinary work, can be procured and worked at considerably less expense than would be involved in using good hardwood. The use of poor hardwood in any work should not be tolerated or thought of under any circumstances, for the simple reason that it is certain to create annoyance and expense, which no house-owner, especially of moderate priced property, should be subjected to.

Pine, however, is not the only soft wood that may be used in house finishing, but it is the most popular, because the most plentiful and doubtless the most satisfactory to deal with. Basswood, poplar, elm, whitewood, spruce and hemlock all make fairly good finishing woods when properly selected and properly treated; but, with the exception of elm, perhaps, all of them require to be either painted or stained before a good finish can be obtained. Elm, black ash and chestnut have all coarse grain, and are unsuited for tasteful work, though all right for ordinary fitments; and indeed may be used for kitchen or bathroom finish in fairly good houses. Later on I will deal with each one of them, and some other of our native woods at length.

The present methods of hardwood finishing are comparatively new, so far as the manner of operation and materials are concerned; as I can find nothing dating farther back than 1850 in the vast amount of literature at my command that treats in any way intelligently on the subject; true, there are methods of varnishing, lacquering and French polishing given; also treatment on finishing in wax, notably a small pamphlet containing a treatise on "The Shining Up of Woodwork," dating as far back as 1809, and published in London. This little treatise is the first to make mention of a wood-filler being employed. The author says that "the pores of the wood ought to be sealed up with a mixture made of ground chalk or rotten stone coloured like the wood, and mixed with glue water. Cover the work to be shined up with the mixture, then wipe off and when hard and dry, smooth off finely with shark-skin and hair-clothe." He then goes on to show how to finish in wax or with shellac, as the case may be. He says nothing of French polishing, but gives directions for varnishing and rubbing down with rotten stone.

To-day there are quite a number of works published on wood-finishing, which shows how much interest is taken in the subject.

The introduction of the modern methods of polishing

finished woodwork has so reduced the cost of fine furniture and elaborately finished woodwork, that now the poor man may have in his house one or more pieces of finely wrought work, as well as the man of wealth. French polishing was a costly operation, and made more costly because of the secrecy surrounding the process. Experts were bound not to teach others the process unless they were well paid for it, and the pupil was compelled to take a solemn oath that he would not divulge the secret or convey to others the method, unless he was paid a large sum, often as high as $100. French polish when well done is certainly a fine finish, and while still costly, is often employed in finishing high-class furniture and woodwork in costly buildings, but its general use is rapidly falling off; yet no wood-finisher is completely rounded off unless he has thorough knowledge of the best French polishing, and in the present work I purpose giving a thorough description of this method, with illustrations of the appliances made use of in the process.

While it is not my intention to write a treatise on the useful woods of America, I do not think it will be out of place to describe a few of them, showing their manner of growth, their peculiarities, durability, and the uses they can be put to, with directions for working and finishing them.

In every description of wood the elementary composition of the organic tissue is the same, but the latter is found associated with very variable organic elements, according to the species of the tree.

Pine trees, for example, contain turpentine, and oak trees tannin. The combustible part of wood is this same organic tissue.

The exterior characteristics of woods are very different from one another. Thus certain wood is soft and tender and of a loose tissue, whilst another is hard and of a compact grain. Thence there is quite a natural division into two classes. The first includes all the soft and tender woods, amongst which may be mentioned the pines, white wood or plane tree, basswood, poplar, etc. The second includes all the heavy and hard wood, such as the oaks, maples, beech, birch, cherry, walnut, etc.

When wood is first cut down as timber, it contains from 40 to 45 per cent of water, but this quantity gradually lessens until it is said to be thoroughly seasoned, when it only contains from 5 to 7 per cent. Moisture, however, is always present in wood, and as it gets older, if exposed to the air, will take in from 10 to 15 per cent. It then becomes very hydrometrical, and loses or absorbs water according to the state of dryness or humidity of the surrounding air.

The density of wood, like that of all porous bodies, can be considered in two different ways, and can be looked for under its apparent volume. The only method which can then be employed consists in forming a block of the wood, the volume of which can be easily measured, and then taking the weight of it. The ratio of this weight to that of the same bulk of water would be the density sought for. This density for the same wood varies according to the degree of seasoning it has had, and to the form and position of the fibers in the sample. A block six inches square cut from a knot, or burl, or a cross-grained part of a tree, would be considerably more dense, and weigh more, than would a block of a similar size cut from the same tree, if it was straight-grained and uniform.

It is the density of the harder woods that makes them popular with wood-finishers. A hard, close-grained wood requires little or no filler, and offers better opportunities to the polisher than do the woods of lesser density.

The use of nearly all kinds of hardwood in the general finish of good houses, has completely taken the place of using pine or other soft woods in wainscoting, floors, doors and general fitments. The variety of the woods employed in these finishings has greatly increased within the last few years. How so many of these beautiful and suitable woods could have fallen into disuse within the present century is one of the decorative mysteries of this period. Walnut, which was the pioneer of the new departure, still remains in use, where cost is no object, although its dull color and unpolished surface are dingy and somber. It has had its use, however, in directing attention to the employment of solid woods, though it is now on the retired list. Mahogany is at present in high favor, the variety known as San Domingo mahogany being especially rare and desirable. When new it has a reddish yellow tone, which grows dark and rich with age. The old wood is much in demand for use in interior finishings and for furniture, and the genuine old pieces are highly prized by their possessors. Cherry is popular for common use, though at first falsified by the red stain, which destroyed its natural beauty and gave it no artificial excellence. Unstained cherry has a yellowish brown color, polishes well, and is excellent for interior woodwork, its moderate cost making it available for general use, though now getting very scarce. If any stain is required, it should be in the dark tones resembling mahogany color.

Black birch is one of the most beautiful native woods, having a grain resembling that of mahogany, and taking a fine satin-like polish. Unstained, it has a warm, rich yellow tone, and its darkened finish can hardly be distinguished from the more costly tropical wood. Yellow birch, though less beautiful and costly, is excellent in grain and color and is often used for fine floorings. The different varieties of oak are well adapted to interior uses, the wood being solid and durable, while each year of wear adds a depth and richess of color and a smoothess and polish to its dark surface. It has a beautiful grain, and is a popular wood for interior finishings and the floors of modern houses. After centuries of wear this useful wood is found in the houses of both humble and royal history as black and smooth as ebony. Butternut resembles oak in color, though not in grain, polishes finely and takes a carved decoration well, and ash and elm are often found as deceptive substitutes for oak, especially in its darker or so-called antique stains. Rosewood is one of the most beautiful woods in use, rich and deep in tone and capable of a mirror-like finish. It is costly enough to keep the merit of rarity.

Apple wood carves finely and polishes well, making excellent panels; so also does the wood of the pear tree. Maple is in high favor, bird's-eye maple being one of the choicest of the light-colored woods, and especially suited to dainty and sumptuous uses. Chestnut and ash are serviceable and inexpensive light woods.

As these woods come into more general use, their genuineness and elegance giving to the houses in which they are placed one of the strongest assurances of that permanency which is the best element of the new

Renaissance, they will be more and more enriched with ornament. And here it will be difficult to avoid the besetting sin of abundant means and inartistic taste, which unfortunately are so often found in company. The honesty of material and the uses which it serves should never be concealed under complex decoration. Most hardwoods are beautiful enough in grain and color to give lasting satisfaction, and for every-day use no decoration beyond a touch of good carving should be applied. Where something more sumptuous is sought, carving and inlaying will make them elegant enough to satisfy the most exacting taste, and when properly varnished or polished no other method of decoration can excel it in beauty or refinement.

I have deemed the foregoing introduction necessary, as it will enable the workman to have some idea of the development of the art of joinery, and the transition from the use of pine and the softer wood to that of the hard wood.

In the following, it is my intention to take each of the woods most in use; describe them, say what I think they are best adapted for, and the best way to finish them, as far as my knowledge extends. By this means the reader will have, under one head, all the book contains concerning the particular wood he is about to finish.

THE BETTER WOODS TO MAKE USE OF

The wood most made use of at the present writing for interior finish, is oak. It is made into doors, sashes, stairs, base, cabinet cases, and wainscot.

The working of oak, particularly in the framing up of panel-work, differs somewhat from ordinary pine or other soft wood panel-work, inasmuch as the latter is generally painted and picked out in two or more colors; thus rendering it imperative to use sound stuff, and to have the work above suspicion. The material is generally so thin that the paneling should be stiffened with stout canvas glued to the back, which is often primed with red-lead paint to afford protection from

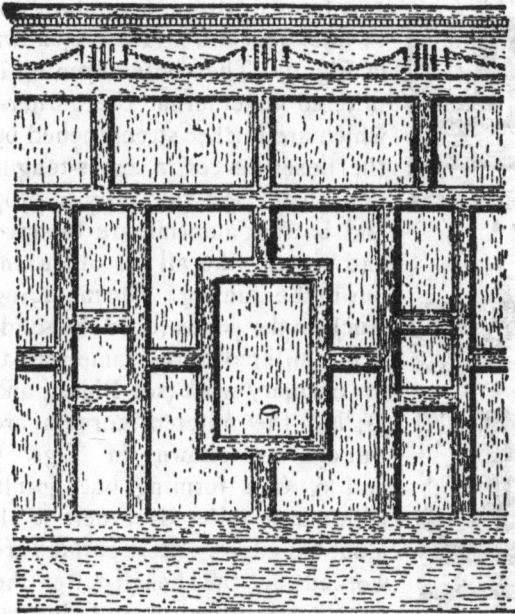

FIG 1   FIG. 2

damp when the work is to be placed in a new building. It is usual to frame up the work with stuff varying in thickness from seven-eighths of an inch to one and a quarter inches, finished sizes. In some workshops it is not considered necessary to clean up the back of the work, though I think it always best to dress down the joints on the back, and true it up with plane and straight-edge. All joints should be close and perfect, and tenons and mortises should be in the center of the

FIG. 3

stuff, and all should be well glued together and left to dry before fixing.

The panel-work shown in Fig. 1 is framed with 1¼ in. stuff, got to thickness and widths as shown in Fig. 2. The center framing is mitered together at the corners, which are further strengthened by the insertion of a cross-tongue joint, while the side rails and top and bottom muntins are tenoned and pinned like all the other tenons in the framing. The moulding stuck on the edges (or in the solid) is not mitered in the same way as for ordinary work (see Figs. 3 and 4); but the mitered corners are worked as shown in Figs. 5 and 6, thus forming butt joints with mason-mitered corners to all moulded edges. These corners are worked on the bench after the paneling is glued up and cleaned off.

FIG. 4

FIG. 5

FIG. 6

The bottom rail is tongued into the skirting as shown in Fig. 8, and the top rail meets the festooned frieze board under the small necking mould as shown, the frieze board being tongued to the dentiled cornice also. This cornice is double-dentiled, one row of dentils being cut farther back than the other, as shown in Figs. 7 and 8. As usual with built-up cornices, this section can be worked on the four-cutter moulder or on a spindle machine; the dentiling, however, being

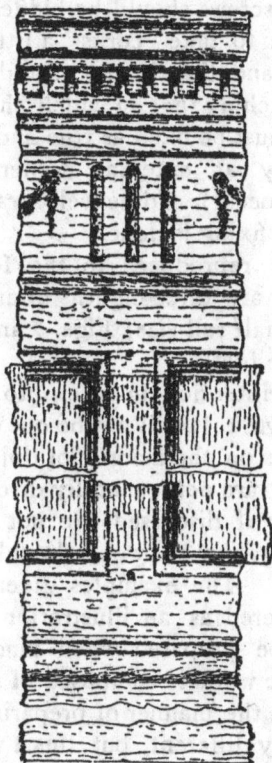

FIG. 7   FIG. 8

best cut by hand. A cover-board lies at the back of the cornice, which is back rebated to receive the front edge of the cover-board. The three flutes over the top muntins have rounded-out top ends, and finish at the bottom on a splay; whilst the festoons are preferably cut out of the solid, but are generally planted on unless otherwise specified.

In fixing this class of work, which is, as a rule, screwed up, all fixing screws should be hidden, or the holes should be bored to take "corks" a little larger than the screw head, and the "corks" should be cut from wood closely matching that in which the hole is bored. The framing must be fixed as true and upright as possible (especially at external corners where mitered vertical joints occur), and be well scraped and cleaned down after the fixing is done.

The illustrations are reproduced to the following scales: Figs. 1 and 2 are three-eighths of an inch to the foot; Figs. 3 to 6, half full size; Figs. 7 and 8, one and a half inches to the foot.

While it is not the intention to enter into the subject of joinery in this work, it will not be out of place to make a few remarks on the manner of doing some odd jobs of work that are not generally discussed in books of this kind, or if discussed, are not done so with a view of showing how the wood should be prepared for the polisher. The examples given in the foregoing are only offered as an illustration of how similar work should be prepared when made from solid wood. Where the work is composed of material "built up" or veneered, the manner of preparing it for the finisher may be very different, but this I will discuss later on.

Suppose a column, or a pillar or spindle is required having a spiral form; unless the workman is well acquainted with the method of laying out spirals, he will be up against a proposition which he will find difficult to solve. To meet a condition of this kind, I show the following method of setting out such work. This example is intended to be prepared for the lathe, but, of course, it may be executed without the aid of a lathe, but in such case much patience and skill will be required.

The sketch Fig. 9 shows the method of setting out the twist or spiral. First turn the wood round; then divide the circumference into four with lines, as shown, marked all the way down. Then, having decided on the size of your twist—and the same details apply to any size and depth —mark the cross-lines, and then the diagonals, which you can easily do with a twist of the leg in the lathe, and both sides as shown by dotted lines. The gouge, held in this direction, will follow the required twist. It is supposed the operator is acquainted with lathe work to some extent, and is aware of the methods and appliances made use of in turning spirals, and the sketch is only intended to instruct him in the manner of laying out the spiral. This will be found useful even in work where the column, etc., is not turned in a lathe, but is wrought by hand.

Handwork of almost every kind in woodwork is rapidly going out of fashion —more the pity—but now and again, men

FIG. 9,

are found who insist on having their work done by hand, and it is well that every joiner should know how to go about the work with intelligence when such work is required. With this idea in view, I give herewith a few instructions and illustrations to help the workman when he is confronted with problems such as I have indicated.

Often dovetailing—an almost lost art—has to be done by hand on some particular job or piece of work, and, in order to prepare the workman for this purpose, I offer the following instructions, and give the accompanying illustration, which I think will materially aid him in his efforts.

Dovetailing to a great many young workmen proves a pitfall, yet, when the method of laying out the work is once understood, it all seems easy enough. A dovetail joint, if made properly, requires no screws or nails, to hold it together or to bring the parts down to a proper joint; but in order to attain perfection there are one or two points which must be attended to when making this kind of joint. First, the stuff must be faced up properly, using the trying plane, on the flat surface. In passing, I will just mention that in dovetailing, whether boxes, drawers, or whatever is being made, the face *side* is always the *inside*, and the face *edge* is the bottom *edge*.

Another point which is often neglected is to put in plenty of "pins" or dovetails. A very good rule for ordinary work in soft wood is to space them not less than 1½ in., or more than 2¼ in., from center to center; for hard wood, or for small work in thin wood, they should be spaced closer still.

In setting out the pins, or rather the mortises in which the pins have to fit, a half dovetail should be placed at both top and bottom, as shown in Fig. 10, and the intermediate "tails" should be brought to an extreme point as shown at B in the drawings—that is, if neatness is an object; but if this has to give way to strength, then "tails" of the shape shown in Fig. 11, should be used.

The bevels of the "tails" should be as shown in the drawings, and in no case should they be shaped as at M, Fig. 14. If made like that, there would be great danger of the sides split- ting off at E, and although this shape at first sight would appear to be stronger than the other, it is really not so in

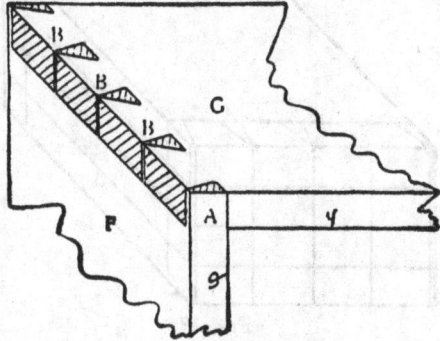

FIG. 10

practice. Also if one, two or three, or even more articles of the same size are being made, the dove- tails need not be set out on one piece only. The whole could be placed together in pairs as shown in

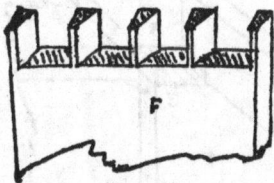

FIG. 11

Fig. 12, and the marks squared over as shown, then screwed in the vise together, and all cut at once with a fine tenon or back-saw, first of all sawing all the pieces off to exactly the same length.

To mark the pins, screw one of the ends upright in the bench vise, with the top end about half an inch above the bench top, as at F, Fig. 13, taking care to keep the face side toward the bench. Then lay one of the sides on it, as at G, so that the squared-over marks on the edges of G coincide with the inside of F, as at H. With the front end of the same saw as was used to cut the "tails," used as shown at I, mark the position of each one on the end grain, and, before removing the side board, number each piece as shown in the figure so that it is known which pieces go together.

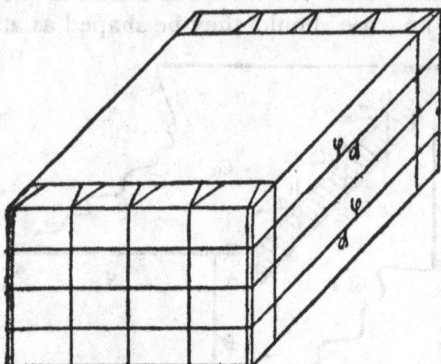

FIG. 12

All four corners of each job in hand must be done in the same way, unless perhaps the work is for drawers or where the front only is required to be dovetailed, although particular notice must be taken that the face *side* of the vertical piece must be towards the bench, and the horizontal piece underneath; and in addition remember that the face edges of both pieces must always come together. These are the edges which

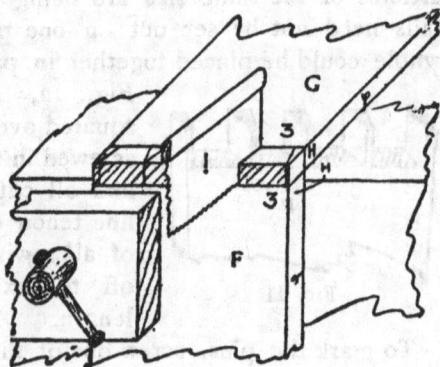

FIG. 13

must be kept level while marking. Neglect of these **points** is the reason of failure to produce good work when making dovetailed joints.

In cutting the "pins" some regard must be paid to the kind of wood being used; soft pine requires more wood left outside the marks than oak or other hard-woods, but tak-ing ordinary work in any of the softer woods, about a sixty - fourth of an inch clear on each side of the

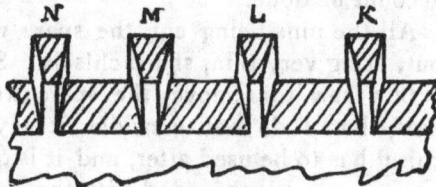

Fig. 14

mark will be ample, leaving rather less at the two half dovetails at the edges.

I should have stated before that in cutting off the stuff for the ends of the box (or whatever is in hand) about half an inch extra length should be allowed, and in squaring across, this extra length should be given to the pin at each end, to be cut when the job is put together.

Before the pins are sawed down, the marks on the outside (that is, where they come to a point) must be squared down as a guide

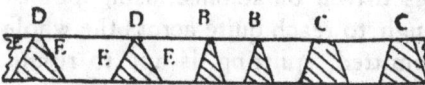

Fig. 15

for sawing parallel with the edges of the boards. This is a somewhat important part, and in Fig. 14 I have endeavored to show how they should and should not be cut. The "pin" K is parallel from point to heel, and this is correct, but not easy to manage. The "pin" at L is cut wedge-shaped, larger at the heel than at the point, which is bad, being liable to split the boards, and also to show a badly fitting joint outside. M is cut slightly smaller at the heel than at the point, which is a good fault—there is no fear of splitting, and, unless overdone, a good fitting joint will result. N is cut out of parallel, which is the worst of all, and must on no account be done.

All the pins being cut, the spare wood must be cut out, using very thin, sharp chisels. Some workers use a bow saw to remove the spare wood between the pins, but I do not consider it any advantage—the chisel has to be used after, and it is quicker to remove all the wood with the chisel at one time.

FIG. 16

I must not forget to emphasize the fact that it is necessary, in setting out, to use knife or chisel for all cross lines, both on the sides and ends. Pencil will not do at all, if good work is expected.

Before putting together, each "pin" must be slightly pointed on all three sides, as shown in Fig. 15, so that they will enter freely, without bruising the wood.

When putting together, the "pins" should be well glued and the sides driven on at once, using a block which is large enough to reach quite across the whole work. If this is omitted, splitting is apt to result.

I have not touched upon putting the work together, as I do not think it necessary, because when the dovetails and "pins" are properly cut, they will be found to go together without any paring or cutting. Indeed, dovetails of any kind requiring fitting before going together may be put down as a botch job and unfit to be further finished.

In this, I have dealt only with simple work, but

there are other kinds of dovetailing, and I present several of them below.

In the figure shown at 17, the end view of a "lap dovetail" is represented, a style which is well known to every joiner who is familiar with drawer making. The method is the same as described as regards the sides, but the ends of the front are gauged on a certain distance, which should, if possible, be the same as the thickness of the sides, and the dovetails are stopped at the gauge mark. The method of marking is the same as before, and the only difference is in cutting the "pin," which has to be done largely with the chisel, as the saw can only be used to start them with.

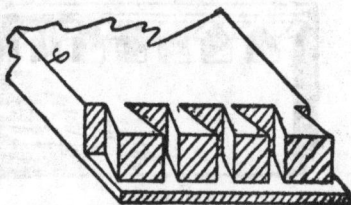

FIG 17

The appearance of the "pins" when cut is shown in Fig. 18, and those who can do the ordinary dovetailing will have no difficulty in making the "lap" dovetail.

FIG 18

In Fig. 19 is shown what is sometimes called "secret," but it is really double lap dovetailing. One part is done as shown in the figure, making the mortises the same size as the pins, and cutting them as shown. The corresponding member is worked as in Fig. 17, the necessary marking being done as in Fig. 13, and marking round point, then cutting them out with saw and chisel. The appearance of this when done is shown in Fig. 20, which is the one marked A, at the side, and Fig. 21, on the top edges; and as will be readily understood, very careful work is necessary in

Secret Lap Dovetailing

FIG. 19

FIG 20

order to make a good job, and, presuming this, the joint is as strong as the ordinary dovetail.

By mitering the top edges, as in Fig. 22, the appearance of this part is as Fig. 23, and if done properly is an improvement. It does not add much to the difficulty of making the joint.

Secret dovetailing proper is a difficult job for any but an experienced man to undertake, but I will deal

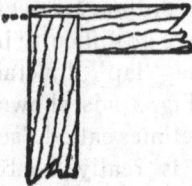

FIG. 21

FIG. 22

with it later on if space permits. It makes a good strong joint if done well, and the appearance is the same as a mitered joint.

Sometimes the end of the drawers, as shown in Fig.

23, is rounded off as shown in Fig. 22, and when such is the case, the drawer projects from the face of the framework as shown by the dotted lines.

In order to enable the workman to better understand the whole theory of dovetailing, I present herewith a couple of sketches which show how secret lap dovetailing may be executed in several ways. Figs. 19 and 24 show how the pins and mortises are laid out when the pins are simply no more than a square tenon.

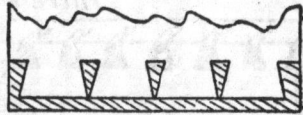

FIG. 23

Secret lap dovetailing is used for a great variety of objects, such as sewing machine covers, instrument cases, etc. Where the dovetails require to be hidden it is not so important that the joint should be at an angle, as the corners can be rounded up to the joint. The difference between secret and plain lap dovetailing is that in the secret method the pins are shortened and the dovetails not cut through, as shown in Figs. 17 and 18, and when put together, in Fig. 22.

FIG. 24

Miter dovetailing, Figs. 25 and 26, is a much more intricate operation, and calls for greater care on the part of the workman. After carefully planing up the wood to a thickness, first gauge on the inside only the thickness of the lap on the end, remembering that the gauge for the lap will also be the gauge by which the ends of the pins and dovetails will be shortened, as shown in Fig. 27, where the dotted lines show the manner in which each piece must be rebated. After rebating, cut the pins, then mark the dovetails and chop them out, after which proceed to miter the laps on both pieces and cut the

Mitre Dovetailing

FIG. 25   FIG. 26

miter across the edge, and, if neatly done, the joint will go together at the first trial.

Bevel or splay dovetailing is much more difficult than any of the preceding, and is not so generally known. The required bevel being given, proceed as follows: First joint the ends to the required bevel, then, without changing the angle, joint the bottom edge to the same bevel as the ends, working from the inside. Now comes the most important part of the operation. The ends must be beveled on the thickness of the wood. This is obtained by laying the blade of the square on the beveled bottom edge, then shooting the ends to the square, working still from the inside of the wood.

FIG. 27

If this part is omitted, the result will be that when the job is put together it will be not only open on the inside but will not be level on the outside—more or less, according to the angle of the sides. Gauge on the thickness of the wood as for plain dovetailing. In cutting the pins, a center line through these should be parallel with the bevel of the bottom edge as shown in Fig. 28, and not cut square to the inside of the job as in plain dovetailing. If this be neglected, and the angle is much, the dovetails will be cut across the grain, and have no strength. After marking the pin with a marking point, before cutting the dovetails, mark them across the ends also, parallel to the bevel of the bottom edge as in Fig. 29,

Bevel Dovetailing

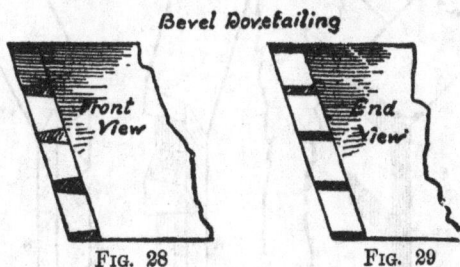

FIG. 28   FIG. 29

and cut accordingly. I would strongly advise the workman who has never made a bevel dovetail to try it upon two pieces of wood, as more will be learned from a little practice when the difficulties have to be surmounted than from any amount of study.

The illustrations shown in Figs. 30 and 31 will convey to the reader a better idea of lap and miter dovetail than the previous ones. Fig. 30 shows the finished corner, the face of the work being shown at F. This may be rounded off if it is so desired. Fig. 31 shows a corner all ready to be driven together, which will have a mitered lap. AA shows both front and side; the shaded laps show the miters.

Another style of bevel dovetailing is shown in Fig. 32, where the pins are at right angles with the line of slope as shown at AA and BB. This style of dovetailing is well adapted for trays, hoppers and similar work. All stuff intended to be used in bevel or splayed work, that is to be dovetailed, must be prepared with butt joints before the dovetails are laid out. Joints of this kind may be made common, lapped or

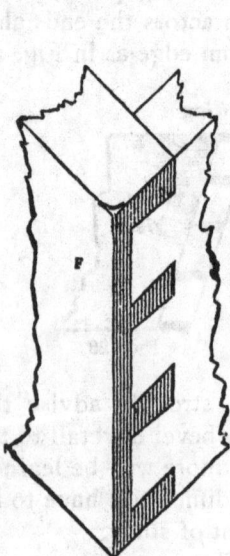 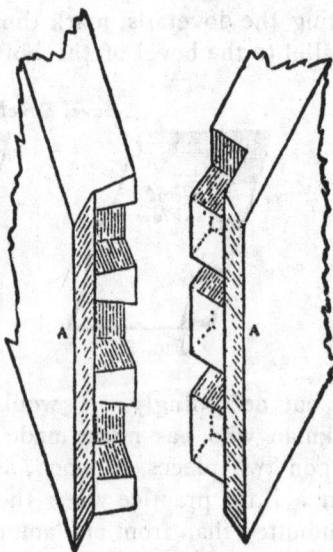

Fig. 30   Fig. 31

mitered. In making the latter, much skill and labor will be required.

In making veneered doors there are a number of different methods practiced by workmen, but it is not in the province of this work to describe them all, or indeed, any of them at great length, but when a number of veneered doors are to be made at one time, the following method, which is adopted in some large factories, may offer some suggestions that may be found useful:

The first operation is to take common coarse white pine boards, with sound knots, and which have been well kiln-dried, the stock used being generally 16 feet long, 1x12 inches, and surface it on both sides by a Daniels planer without regard to thickness. Some boards are thinner than others, while others are warped in drying, and the thickness of the boards is

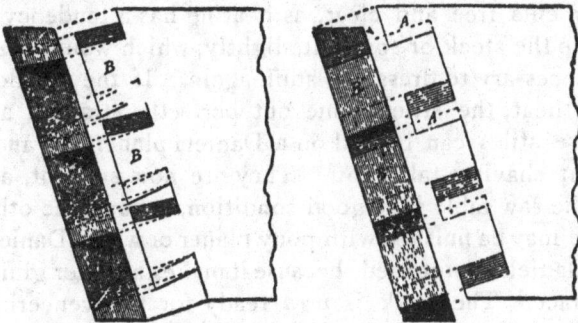

Fig. 32

immaterial, perfectly seamed surfaces only being necessary. After the stock is planed it is cut into such lengths as the bill of doors calls for. They are ready now to be glued up. The face board, of what ever hardwood is to be used, is planed generally to ¾ inch thick, and is also run through a Daniels planer. The stock is now ready to go to the gluing press, and as the Daniels planer makes the best gluing surface in the world, no scratch planing is needed. After properly heating in a box the stock is brought out and carefully glued, the hardwood face parts being marked for it. From three to five parts are put in the press at one time, and a pressure of 20 tons, brought down by screws, is put upon these parts. After remaining in the press the proper time they are taken out, and generally remain several days before being worked up, which gives the glue plenty of time to harden. When ready to work again these parts are taken to a Daniels planer and squared up, after which the parts are taken to a very nice cutting table or bench saw, and are cut up to sizes required, leaving them ⅛ inch large for future dressing. It is a positive necessity that the saw cuts free and clear, as heating has a tendency to warp the stock or spring it slightly, which would make it necessary to dress the stuff again. If the saw does not heat, the stiles come out perfectly straight, and these stiles can be laid on a Daniels planer bed and a light shaving taken off. They are now straight, and if the saw table is in good condition, square; the other side may be finished with pony planer or with a Daniels. A Daniels is preferred, because it makes a better gluing surface. The work is now ready for the veneering, the thickness of which is immaterial, as it may vary from the thickness of thin paper to ¼ inch. Heated cauls are now used for the veneer, and the stiles, if heated at all, are just warmed and the veneer glued on by piling up with a hot caul between each stile. The old-fashioned way of making veneered doors may do very well when only two or three doors are to be made, but in these days of sharp competition, manufacturers are obliged to adopt the quickest methods compatible with efficiency and good finish.

The doors or other work—for this method will apply to wainscot or any similar work—should be placed in a dry room for a day or two, when they may be finished up and made ready for the polisher or varnisher.

If first-class doors are required, it is always better to build up the stuff for the stiles and rails, and glue them together before they are veneered. A white pine door is about the only door that can be made successfully from solid wood. In a house with, say, a dozen doors, what other wood is there that will absolutely hold its place during a reasonable period? Certainly yellow pine will not do it. A solid oak door is a pest, and should not be put in a house except under written instructions. Sycamore cannot be used solid, and certainly neither gum nor maple. Possibly walnut or butternut might, but who would think of using them under present conditions?

The door shown in Fig. 33 is an illustration of one of the most serviceable doors made. The section of stile shown in Fig. 34 will give an idea of the

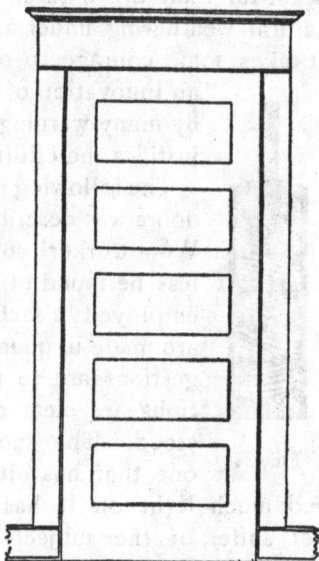

FIG. 33

manner in which the stiles and rails are built up; the veneer showing on the outside, also the panel.

Work of this kind may be used outside as well as for inside, and no one need fear unpainted white pine or oak for outside work. There is in Europe abundance of woodwork, exterior and interior, that has not been tickled by a brush in over five hundred years. All the native Swiss cottages are unpainted on the exterior.

All of the half-timber houses of France, no matter how richly decorated by the great artists of the chisel, are unpainted, and we have known of work in white pine and quartered white oak without the use of paint which has been in existence for centuries untouched by paint, and time justified the experiment. Nothing is more beautiful than the natural grain of the wood and its natural weathering under a proper transparent finish. It takes some courage to organize and push through an innovation of this kind, and one is beset by many warnings, but the beautiful result justifies the effort.

The following method of making veneered doors was described by H. T. Gates in "The Woodworker" some time ago. It will doubtless be found of much benefit to workmen employed in factories where veneered doors are made in quantities. The hints and suggestions are to the point, and the instructions are clear and may be readily understood. The subject of veneered doors is one that has often been discussed of late and much light on it has been thrown, yet it will not suffer, if the subject is again brought forward, inasmuch as the various factories have particular ways of their own of producing this rapidly growing popularity in the door line. Instead of trying to lay down any fixed rule, saying thus and so shall they be made, let us consider some of the essential points which may be adapted to suit each condition as the manufacturer finds it.

FIG. 34

I. **Equipment**.—Aside from the usual door-making machinery, this consists of larger facilities for preparing and applying glue, veneer press, resaw for veneers and panels (unless they are purchased from a dealer in veneers), a warm room where the glue may be applied and material gotten ready for the press. Unless the factory is already supplied with a large kettle for preparing glue, it will be found of advantage to make a large copper kettle that will fit the holes in the heater, but large enough to hold three or four ordinary-sized kettles of liquid glue. This can be done by making it higher and wider about the flange, as shown in Fig. 35. In this

Fig. 35

way sufficient glue may be made ready for a good-sized batch of doors without fear of running out.

Of course, these remarks do not apply to the factory having modern glue-spreaders, hand or power-feed, which are very essential in strictly veneer establishments. Our remarks apply only to the shop where veneered doors are one of the many by-products, so to speak, which accompany the usual line of mill work.

For spreading the glue by hand, procure a 4-inch flat wall brush and prepare it by pouring alcohol gum-shellac into the roots of the bristles, and driving blind staples into the butt end, as close together as possible, thus preventing the bristles from coming out. Have a pair of "horses" about 3 feet high, strongly made, and having an angular piece on the top, to prevent waste of the glue, and squeezing it off the work at the bearing point—see Fig. 36.

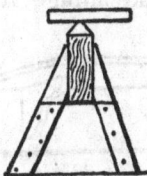

Fig. 36

Make the veneer press wide enough between the uprights to permit of veneering a table top or wide panel if needed, and have two rows of screws, so that two stacks of cores may be pressed at one time. A strong press can be made with 6x6-inch maple cross-pieces and ⅞-inch round iron rods, with jambnuts to hold the upper piece in place, having regular veneer press screws at least 1⅜ inches diameter. The bedpiece should be lagged up and trued, so that it will be straight and out of wind. To prevent the work from sticking to the bed, it should be covered on the top with zinc or tin—see Fig. 37.

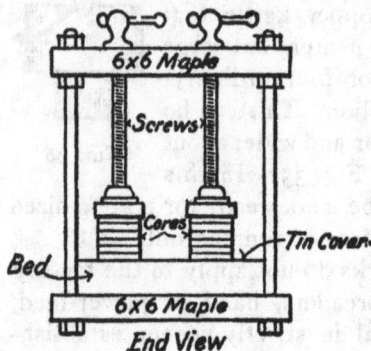

FIG. 37

For resawing veneers and panels, where there is no band resaw, a gauge (Fig. 38) and pressure roller (Fig. 39) used on the band saw table, and 1½-inch saw in proper trim, with right management should turn three or four veneers out of inch stock, which may be applied direct to the coring without dressing, as described later on.

To do a good job of gluing to advantage it is necessary to have a warm room, so that a large batch of material may be worked at one time. There are various other purposes to which this room may be put, but

FIG. 38

to make a veneered door properly, without waste material or lost time, a warming room is very essential. First, the parts of wood to be glued must be thoroughly warm; also the temperature of the room where the work is to be done must be such as not to chill the glue and hinder its spreading and making good joints.

It is taken for granted that our factory has a hand jointer, pony planer, mortiser, tenoner, sticker, door clamp and drum sand-

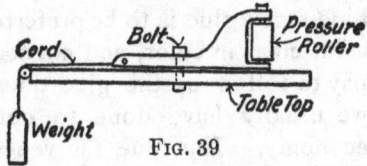

FIG. 39

er. It is a question whether veneered doors can be profitably made without the few essentials here enumerated, and where they are made in quantities, special machinery for spreading glue, cutting mouldings, presses, clamps, panel-raisers, etc., will be needed.

II. Materials.—Dry coring is the first thing that is required to make good doors. It is usual to cut up the material and put it in the dry (or warm) room referred to above, or in a dry-kiln, properly stacked, and leave it there as long as possible to drive out every particle of dampness. All waste material of suitable size and too poor to be used for any other purpose may be used for coring. It is preferable to have the strips wide enough so that when glued up they may be split through the center to make two stiles or rails—see Fig. 40—thereby saving much labor in gluing, which item cuts quite a figure in veneered door work. The stock sawyer

FIG. 40

can lay aside such material from time to time and have it stored as mentioned, so that there is a supply of dry stuff to draw on when a batch of doors is wanted.

The glue for coring need not be of high grade, and where quick preparation is desired, a ground or pulverized bone glue will answer the purpose admi-

rably. The men soon become accustomed to handling
the glue and it needs no soaking, as the flake or
noodle glue does. For veneering a medium grade of
hide-stock glue is to be preferred; one that is free from
acid, clear in color, and not too quick-setting. It will
pay to follow up the glue question more closely than
we usually have done to obtain good results with
economy. The glue for veneering does not want to
be too thick. Practice in the work makes the workman proficient in its preparation. It should flow freely
from the brush without being "tacky," as the painter
would say.

The veneers and panels should be cut up and
resawed before they are kiln-dried. The ends should
be glue-sized, and they should be stacked straight and
even in the kiln. Those who have tried resawing kiln-
dried hardwoods, are aware of what a sorry job it
makes; and how the veneers buckle, spring out of
shape, pinch the saw or make it run crooked. When
the saw has not too much set, the veneers may be
glued onto the cores without planing, provided the
sawing is a good, smooth job. Care must be taken
in dressing veneers or panels, not to chip them out, as
that is ruinous in this class of work.

III. Construction.—The man who is doing this
work needs to be familiar with the work and its
methods to do it well and economically. Filling the
doors is the first work towards the desired end. A
list of the size, style, thickness of doors and kinds of
wood should be on each working bill, and follow the
material in its progress out of chaos into stiles, rails,
panels and finally the finished product. This bill
should include the edge strips, the width, length and
thickness of each bundle of cores. the finished size of

the parts they are intended to make, and the number and both the sawing and finished size of veneers and panels.

After the sawyer has the material cut, and it is thoroughly dry, the one who does the gluing assembles the cores, puts them on the heating coil and prepares his core glue; the pieces are spread on the horses and given a coating of glue, assembled in batches, and put into the press, the surplus glue being squeezed out by this process, which includes putting the edge strips on each stile requiring one.

After they have been allowed to set sufficiently, they are taken to the jointer and the straightest side trued up. If they are built-up for making two pieces, they are resawed and again jointed and thickened to desired size on the pony planer.

They are now ready for veneering. They are again put in the warm room over the coils; when warm, they are put on the horses as before, and spread with glue on both sides; a bottom board is first laid and then the veneers and cores stacked in regular order. The veneers must previously be carefully looked over, poor ones culled out, and any pin holes, porous spots or checks covered by gluing a piece of paper over, to prevent two stiles from being stuck together by glue oozing through such spots. They are again pressed out, and when dry, trued and sized to width. They are now ready to be laid out, same as any blind-tenon door.

The framing must be done in a first-class manner, with true joints and tight tenons. In fact, all machine work on veneered doors must be carefully done to have true work and tight joints.

Instead of putting the panels in when the doors are

put in the clamps, the framework is glued together
with open panels, the stiles and rails being grooved,
and after the doors are polished and put on the
finishing bench, a panel strip is put in all around the
edge of each panel, to which the panel mould is glued
and nailed.   The moulding is put in one side first,

panels laid in, and moulded the
other side, as shown in Fig. 41.
This arrangement prevents the
moulding from pulling away
from the stiles, should the pan-
els shrink, and allows enough play for the panels to
keep straight with the natural working of the wood
in the changes of the atmosphere. There is advantage,
too, in gluing up the framework without the panels.
This cannot be done in the case of solid moulded doors.

FIG. 41

The finish of a veneered door should be first-class;
the panels, moulds and framework well sandpapered,
and flat surfaces scraped smooth, as every defect seems
magnified when the filler and varnish are applied.

Special care should be taken not to scrape, scratch
or mar the face of the doors in shipping. Many a
good door has been injured by careless packing or
handling in shipping, after the cabinetmaker has
finished his job. They should be crated, if shipped
on a railroad or by boat, or they will not be worth
much on arrival at their destination.

Wedged Doors.—The day of the wedged door has
passed, and all modern-built houses contain what is
known to the trade as "blind-tenon doors." The
"dowel" door is practically a blind-tenon door. In
plants where a set of dowel door machinery has not
been installed, the problem of making these doors
presents itself.

The advantages of this door are the saving of lumber on the rails, of time in laying out all stiles both sides and mortising them from both sides, the neat appearance of the stiles, especially on natural-finished work, and the ease with which they may be glued together.

Several points must be kept in mind in order to secure success. Let the stock sawyer cut all rails

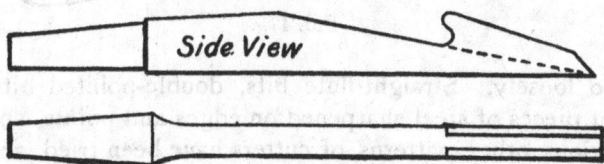

FIG. 42

exact, so the tenon will not touch the bottom of mortise before the shoulder is tight at the coping. The tenons should fit more snugly both sides and endways than in the old way, to hold well and make tight joints on the muntins. The glue should be applied to the mortise in such a way that it reaches the tenon, as well as the shoulders of rails and stiles, to make a strong job of the gluing.

The cleaning out of chips from the mortise has been a difficult problem, and it is not yet solved satisfactorily. Chain-saw mortises obviate this, but they have their faults, too. The "lip" chisels, Fig. 42, clear out the loose chips, but leave the fine chips that are pounded down by the action of the chisel, to be removed. In order to do this, a long S-shaped chisel with large wooden handle, Fig. 43, is used. The blade is ¼-inch wide, and tapers from o at the cutting edge to ¾ or ⅞-inch at the handle.

This is a slow, laborious job. An easier method is

to use a twist drill bit in a horizontal boring machine, leaving the arbor in a fixed position and moving the stiles back and forth, both lengthwise and sidewise, by hand. The bit should not be quite as large as the mortise, so as not to enlarge it and make the tenons fit

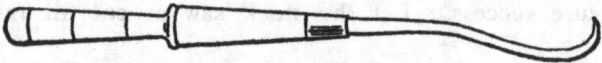

FIG. 43

too loosely. Straight-flute bits, double-pointed bits, flat pieces of steel sharpened on edges and points, and various other patterns of cutters have been tried, and drills give the best satisfaction; only, they are easily broken.

The doors should be framed and stood in a hot room for about a week, to dry out. When ready to glue together, they are warm and do not chill the glue as it is applied. The kind of glue has much to do with the rapidity with which doors may be glued up. Expensive glue is not required for this work, but a strong, *quick-setting* glue is needed, so that the doors do not have to remain too long in the clamp, thus retarding the work. A light-colored glue, having had a sufficient quantity of oxide of zinc mixed with it by the glue manufacturer, is the kind required. Using this, the man at the door clamp can take them out of the clamps about as fast as he can properly apply the glue, without their springing apart at the joints so as to require small clamps to hold them.

Bear in mind, in making blind-tenon doors, you must have good machine work, dry lumber, snug fit to tenons, quick-setting glue, all applied in a good, sensible manner. The kinds of wood and styles of doors both affect the results obtained, and must be made the subject of study in order to succeed.

A good tool for rubbing out the surplus glue after the veneering has been put in place is shown in Fig. 44. The rubbing part may be made out of an old plane iron, or from a worn-out scraper.

THE CHOICE AND USE OF GLUE

Glue is an article which plays an important part in carpentry. It is therefore to the interest of all to know how to choose a good article, and also how to use it properly for various kinds of work.

First, as to how to recognize a good article. It is really useless to recommend Russian, Scotch, French, etc., as there are good and bad qualities of each, and we make as good glue in this country as anywhere in the world. Neither does the test of looking through the cakes at the light, and choosing those of only a bright color, apply in all cases, although it is a fairly good test with thin glue, that is, glue which is sold in thin cakes. The best test which the writer has found is to break a cake into two or three pieces, either by forcibly bending it, or by striking with the hammer. If the fractures present a smooth, even surface, the glue is poor in quality; but if, on the contrary, they present a ragged appearance, leaving any amount of sharp splinters, it is evidence of good quality, and may

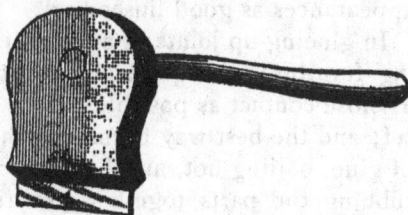

FIG. 44

be depended upon. The latter is to be preferred, as being best able to stand the damp weather without going mouldy and perishing.

The best way to prepare glue for use is to break up sufficient to fill the inner vessel of the glue pot into small pieces, and fill the vessel with water. After soaking a few hours, the outer vessel can be filled with water, and allowed to boil. When this takes place, give the contents of the inner vessel a good stir occasionally until all is melted, when the glue should run off the brush freely, and be almost as thin to all appearances as good linseed oil.

In glueing up joints of any description, or in repairing furniture, the great object is to get the wood into as close contact as possible before the glue begins to set; and the best way to manage this is to put on plenty of glue, boiling hot, and by means of pressure or by rubbing the parts together, to rub out as much as possible. The general idea is that a certain amount of glue must be left in a joint, and that it will hold best if it is reasonably thick. Greater mistakes were never made. The thinner the better, and the less left in the stronger the joint will be; therefore it follows that unless the parts to be joined fit closely together, a strong joint is impossible.

The glue pot must be kept free from dust and damp, as glue which goes mouldy, or gets mixed with foreign matter, is useless; therefore, unless it is used fairly often, only a moderate quantity should be made at one time.

The outer or water vessel of the glue pot should be large, so as to be capable of holding enough water to keep the glue hot long enough for any reasonable job. A small water vessel is a continual nuisance, owing to its continually boiling dry when making glue, and failing to hold the heat when using it. Therefore, when purchasing, do not get one which holds less than a quart of water at the least. The extra cost will be saved in a very short time.

For large establishments where much glue is required, special appliances are provided, and may be obtained from most dealers in plumbers' goods.

For glue to be properly effective it requires to penetrate the pores of the wood, and the more a body of glue penetrates the wood, the more substantial the joint will remain. Glues that take the longest to dry are to be preferred to those that dry quickly, the slow-drying glues being always the strongest, other things being equal. For general use no method gives such good results as the following: Break the glue up small, put into an iron kettle, cover the glue with water and allow it to soak twelve hours; after soaking, boil until done. Then pour it into an airtight box, leave the cover off till cold, then cover up tight. As glue is required, cut out a portion and melt in the usual way. Expose no more of the made glue to the atmosphere for any length of time than is necessary, as the atmosphere is very destructive to made glue. Never heat made glue in a pot that is subjected to the direct heat of the fire or a lamp. All such methods of heating glue cannot be condemned in terms too severe. Do not use thick glue for joints or veneering. In all cases work it well into the wood, in a similar manner to what painters do with paint. Glue both surfaces of your work, excepting in case of veneering. Never glue upon hot wood, as the hot wood will absorb all the water in the glue too suddenly, and leave only a very little residue, with no adhesive power in it.

REMARKS ON VENEERING

A wood suitable for veneering requires to be thoroughly well seasoned, free from knots and shakes, and should not contain turps. The best of woods for the purpose are mahogany and American walnut, although good pine answers well for ordinary purposes. The surface, if flat, is carefully planed with the trying plane. It is then well toothed over with the toothing plane—first the lengthways of the wood and afterward the crossways—care being taken to tooth the work thoroughly. If you are working pine, use a coarse toothing iron; if mahogany or other hardwood, a finer iron is requisite. If the wood presents a hollow or rounded surface, it is shaped with suitable planes, rasps and files, and finally well prepared crossways with coarse glass paper such as strong No. 2 or No. 2½.

The next preparation is sizing. To make the size, take one part of good glue and boil it well with 50 parts of water; then brush over the ground-work while hot; allow to dry, and, if there should be any defects in the ground-work, fill in with stopping. Make your stopping by mixing some finely ground plaster of Paris with hot glue and water, enough to form a moderately stiff paste. Then lay in where necessary with a chisel, taking care to allow for shrinkage; let it dry, then level off with a rasp.

Having sized the ground-work over, next proceed to the preparation of the veneer while it is drying. Look carefully to the wood before cutting it, and see that it is done in such a manner as to get the grain of it to the best advantage. Cut it rather larger than the surface it is intended to veneer, to allow for leveling at the ends and sides. Most veneers, such as mahogany, oak, chestnut, maple, sycamore, birch, satinwood and various other woods, are ready for cutting as received from the merchant; but some, like burr walnut, brown oak, Amboyna, etc., present an uneven surface, called "backly." When this is the case, damp one side with clear water, lay it down with its dry side upward, and put the wet side of the next veneer upon it, repeating the operation till all are done. Take particular care to keep each veneer, if there are more than three or four, in its proper order as you damp and turn over, and do not on any account get them mixed. Let the wood stand about four or five hours, then lay them out to allow to nearly dry and they will be ready for cutting out.

The next process is flatting. Get two pieces of wood (dry, straight pine will answer), rather longer than the veneer, and heat them on a stove or before a bright fire; then place the veneer together between the hand screw, and allow it to remain for about half an hour; repeat this operation until the veneer is perfectly dry and thoroughly flat.

Our wood is now ready for filling in. If it is perfectly sound this operation is, of course, unnecessary; but it frequently happens, especially with burr walnut, that it contains holes that require filling. To do this, take a piece of the veneer (off the edges of that already cut out), and flat it precisely as the other. Select the part of it which matches best with the grain of the wood around the hole to be filled in. Place this underneath the hole. If you have a stamp rather larger than the hole, you may now cut it square or circular and the piece for filling it at the same time. If not, take an ordinary pocket knife having a sharp point and cut your hole and veneer the required shape.

When you have filled in the wood, lay it on a flat board, then press the piece in with a hammer. If they are rather large use one or two finely pointed tacks to keep them in position. Now cover all the pieces with strips of paper, selecting a strong paper such as copy book or note paper for the purpose—one that is not too thick—and glue it on one side. Take care to use glue just thick enough to hold the wood in position. Pay particular attention to this, or it will cause a good deal of trouble. You will find it best to cut the paper in strips about 1¼ to 1½ inches wide. Lay it on a board to glue and smooth it over your veneer with a damp rag.

Jointing.—We now proceed to jointing. Place the veneer in the position it will appear when laid. Observe that it matches. If you are to have one joint with two veneers or two joints with four veneers, see that the grain of the wood forms a figure having both sides alike. If the veneers have b en kept in their right or following order, this will not be difficult. If you are working a thick veneer (saw cut), make the joints with an iron plane or ordinary trying plane on the shooting board; if using thin veneer (knife cut), make them with a chisel and straight-edge. Take particular care to have the bevel edge of the chisel against the straight-edge when cutting or it will run, and you may come off with an ugly cut. Now put the jointed edges together on a deal board, and tack one edge down; put the tacks about ¾ inch from the jointed edges and about 2½ inches apart. Having tacked one piece down, put the other up to it and tack it in the same manner. Now cover all the joints with paper, glue together in the same manner as previously mentioned in the filling in; smooth it well down with a damp rag and allow to dry. If the weather is hot it is best to cover your joints to prevent them drying too quickly. A good and simple method is to lay your board with the veneer downward on the floor. Let the joints dry, then take out the tacks and knock the head holes in with a hammer from the underneath side. Put the veneer aside until you are ready for laying it. It is best to cover it up and keep the air from it by placing it under a board or wrapper.

There are two ways of laying veneer—by means of a caul or a veneering hammer. I shall describe both methods, although the first is of greater importance, and should, whenever practicable, be adopted, but in certain cases which I shall mention the second is extremely useful.

Veneering by Caul.—First make a caul, then take a piece of well-seasoned cedar or pine, rather larger than the surface intended to cover (about 1 inch to 1½ inches each way), and plane it up true on both sides, if the work is flat. If otherwise, make it to the requisite shape to. fit the work, hollow, round, or whatever it may be. If it is necessary to shape the caul, use thicker stuff, and it is advisable to screw on two or three battens on the back. When making shaped cauls it is best at the same time to get out the pieces of wood necessary to form a flat surface when the wood is put in the caul. Thus, suppose we wanted to veneer a door having a rounded surface on one side and a hollow one on the other. We have made a hollow caul to correspond with the rounded surface having its under side flat. Now put the rounded side of the door in the caul and shape your piece of wood, rounding it to fit the hollow side. They should be 2 inches wide—the same width or a little larger than the caul, and 5 inches or 6 inches apart. If one side only of tne wood is shaped these woods are unnecessary. If the wood is not wide enough, make a good joint, dowel it together, and take it to pieces for heating. The caul, if likely to be much in use, should be covered with zinc. Cut the metal out large enough to cover the face of it, with sufficient to turn over the edges and ends, and fasten it on with flat-headed zinc or copper nails.

Numerous failures in unaccustomed hands may be ascribed to bad glue. Nothing but the very best glue should be used for veneering. Get the very best glue you can, break it up and boil thoroughly. It differs so much in strength that the proportion of water cannot be given, but after breaking up in pieces just cover with water and allow it to soak; then boil off with frequent stirring. It will, if good, now require about half as much water as previously added to bring it to the right consistence for veneering. It should be spread evenly with the brush and be free from lumps. Having made the caul and prepared the glue, get the hand screws and cramps to commence laying. Heat the caul on a stove or before a bright fire. If it is to be doweled together, and if it is more convenient, take it to pieces, taking care to mark your joints first. If you intend laying two similar pieces of veneer on flat surfaces, heat both sides of is and do both pieces together. If not, get one side of it well heated, as hot as you can without letting it burn. While it is heating, set the hand screws and cramps open as near the distance as you will require them, and place handy for the work. Now glue the ground-work well, and if the veneer shows any signs of being backly glue it slightly on the underneath side, as this will help to soften it. Having finished gluing, put the veneer on the work and smooth it over gently with the hand. Then see that the caul is hot enough, and that its surface is free from any small cinders or dirt. Now rub it over with a greasy rag, and lay it gently on your veneer. Draw the work and caul a little over the edge of the bench, just enough to get the hand screw on; put it on very gently, then tighten as much as possible. You can then stand it upon the floor, and if you have nobody to hold it for you rest the hand screw against the bench while you put on the remainder. They should be placed about 6 inches apart, and mind that they bite fairly. Do not get any screws tighter than the other, or you will only get the pressure at the outside and inside of the cramps. If you have a piece of work so wide that the screws will not reach the center from either side or the ends, get two pieces of wood 2 inches or 2½ inches thick and about the same thickness, plane them up, slightly rounding on one side, put their rounding sides facing each other on the work, and hand screw them at each end; they will then tighten in the middle and give sufficient pressure. Let the caul remain on for nearly an hour (in very hot weather longer will not hurt), then undo the hand screws, and if the caul sticks, insert the edge of a thin metal square, the back of a hand saw, or anything of a similar nature, between it and the veneer, and work it carefully about until you get them apart. If the glue has been used thick enough and the caul well greased, there will not be very much trouble, and they will often come apart themselves, or by giving the end of the caul a tap with the hammer, or on the end of the bench. See that the veneer is down. Feel it all over with the hand. If it is up you will be able to tell by the hollow sound on tapping it with the tip of your finger as well as by the raised appearance called blisters it will present when held to the light. If you heat your caul sufficiently, use the glue thick enough, and put the hand screws on properly, you will not be troubled with blisters. Should, however, there be any, let the work stand for one or two hours, and then put a smaller hot caul on when required until well down all over.

To level the veneer, first lay it (veneer downward) on a board and scrape off with a chisel as much of the glue that has come over the edges as you can. Now put it on the bench screw, and level toward you, with the paring chisel if thin veneers; if thick, use a smoothing plane. Put it aside to dry. If you have two pieces of the same size put them with their veneer sides together; if only one, place it so that the air does not get to the veneer; allow to stand for two or three days, then scrape off the paper for filling in pieces and jointing with a chisel, having previously damped it with hot water. The work is now ready for sizing. This operation may be dispensed with, but it is decidedly advantageous, especially if working wood which has an open grain. The size, which should be about the same strength as that used for the ground-work, is brushed or rubbed over the veneer with the hand, then wiped off as dry as possible with a cloth.

As has already been mentioned, this method is useful in certain cases. We sometimes want to veneer an edge, to put a narrow slip of veneer on some small surface where it would be very inconvenient to caul it down. If you are working a wood of a glossy or greasy description, like satinwood or rosewood, its nature will not admit of sufficient pressure by this process, nor should it ever be adopted for work where water will act injuriously. I believe the prevailing opinion is that veneer requires a good deal of water to make it lie. In the first process, you will remember that it is laid quite dry. In the second process water is used, and if we consider that a damp surface tends to cause the wood to cast as it dries, we can readily understand where it should be used.

Veneering Round and Tapering Columns.—Occasions often arise where it is desirable to veneer columns of varying size and shape, and just how to do the work to the best advantage is a question not always readily determined by the workman.

In commenting upon this matter a well-known authority offers the following suggestions concerning tapering columns:

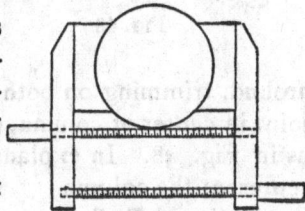

Fig. 45

"The success of this work depends entirely on the proper cauls. The sponging and gluing is the same as the work previously mentioned. Take a circular column 3 or 4 inches in diameter, the core to be made of any suitable material. Pass a piece of paper around the core and make a straight joint through the center and we have the exact size of the veneer. A caul of tin with a cleat on each end for the purchase of the hand screws is to be made as in Fig. 45, leaving the top open for an inch or more to give the glue a chance to escape. Apply the glue to the core and pass the veneer around it, not forgetting, however, to sponge the veneer before doing so. After screwing together, put the whole in the steam box to warm up the stiffened glue. After the work has become thoroughly warm, take it out and give the hand screws a few turns, then lay it aside until the glue is set. Take off the hand screws and dispense with the tin cauls. The ends may then be brought together, as shown in Fig. 46, only the caul on the joint is to be heated.

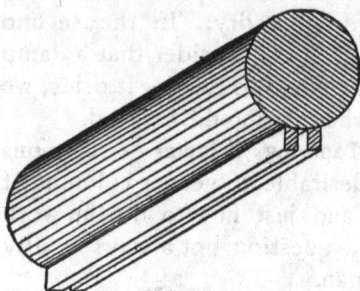

"Fig. 47 is a tapering column. The work is the same as above described, only the shape of the veneer is different, which may be got either by passing paper around, trimming on both ends and making a straight joint in center of column, or by laying out the diagram, as in Fig. 48. In explanation thereof, let A, B, C, D represent the column in question. To find the shape of a veneer to pass around this column extend AB and CD to E. With the radius EC and ED, describe the arcs CF and OH, which will be the required shape, and the distance AF and IH will pass half-way around, and the veneer requires to be cut as large again to pass all around the column. Under no circumstances make more than one joint. By following the instruc-

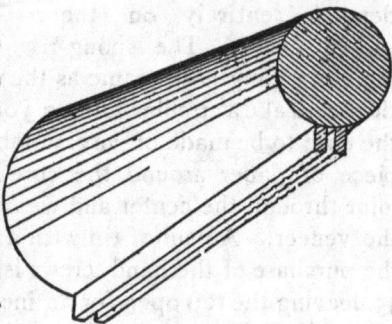

FIG. 46

FIG. 47

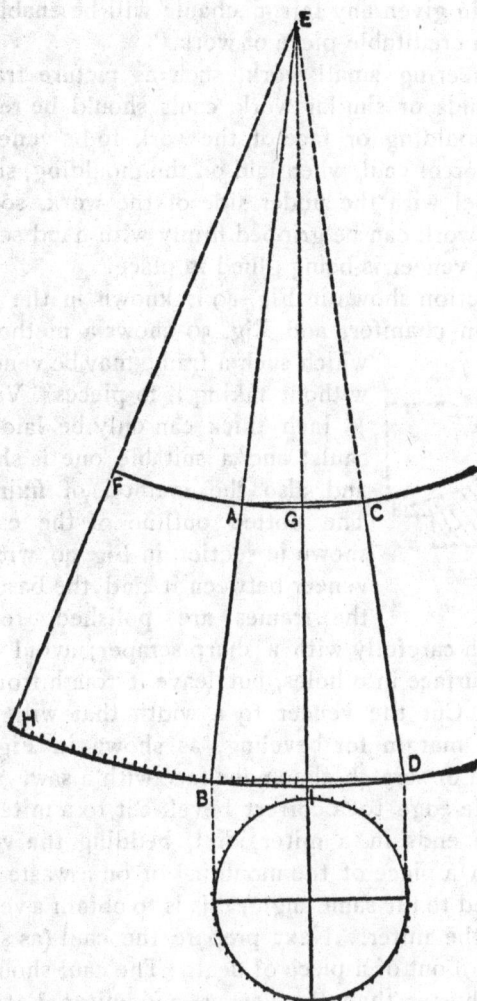

Fig. 48

tion herein given any fair mechanic will be enabled to execute a creditable piece of work.''

In veneering small work, such as picture frames, clock stands or similar work, cauls should be reverse of the moulding or face of the work to be veneered, and the top of caul, when laid on the moulding, should be parallel with the under side of the work, so that caul and work can be gripped firmly with hand screws, when the veneer is being glued in place.

The section shown in Fig. 49 is known in the trade as a plain chamfer, and Fig. 50 shows a method by which such a frame may be veneered without taking it to pieces. Veneer ⅛ inch thick can only be laid with cauls; and a suitable one is shown, and also the method of fixing it. The dotted outline of the caul is shown in section in Fig. 49, with the veneer between it and the base. If the frames are polished, remove the polish carefully with a sharp scraper; avoid working the surface into holes, but leave it rough from the scraper. Cut the veneer to a width that will allow sufficient margin for beveling, as shown in Fig. 49, and tooth or scratch the under face with a saw. Joint the inside edge to a correct bevel, cut to a miter and joint the ends in a miter joint, bedding the veneer either on a piece of the moulding or on a waste piece chamfered to the same angle; this is to obtain a vertical face to the miter. Next prepare the caul (as shown in Fig. 49) out of a piece of deal. The caul should be ¼ inch shorter than the veneer, and mitered at each end; not cut square as shown in Fig. 50; which is drawn thus for clearness. Make the caul slightly

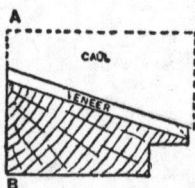

FIG. 49

round in length and beveled, so that the two surfaces marked AB (Fig. 49) are parallel. Well glue the veneer and the frame, lay the veneer in the proper position, place a strip of paper on the surface of the veneer, and lay the caul (preferably hot) upon this paper and fix with hand screws as shown, the more screws the better. Take care that the inside edge of

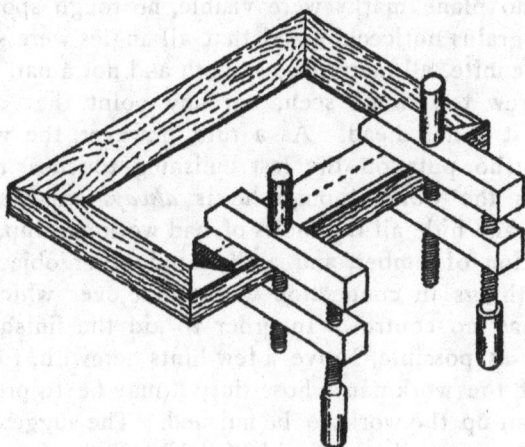

Fig. 50

the veneer is flush with the edges of the frame. Proceed to fix the opposite side in like manner; then wipe off the superfluous glue with a rag dipped in hot water, and wash the miters clean; then fit in the two end pieces, which can be fixed in the manner described for the sides. Allow twelve hours to dry, then clean off the back edges and scrape up the faces.

MAKING READY FOR POLISHING

If the workman who finally finishes woodwork had the selection of the materials out of which the work

is formed, and wrought the work himself, the following suggestions would scarcely be required, for he
would see to it that the material was free from shakes,
cracks, worm holes, doze, sap and fractures; and he
would so choose his material that the full beauty of
the grain would show, when the polish coat went on.
He would also see that the work was *clean;* that is,
that no plane marks were visible, no rough spots or
cross-grains noticeable, and that all angles were sharp
and definite, all mouldings smooth and not a nail head
or screw top to be seen, nor any point that could
suggest a nail head. As a rule, however, the workman who puts on the last finishing touches never
makes the work, though he is *always* expected to
cover and hide all the faults of bad workmanship, bad
selection of timber, and a thousand other objection-
able things in connection with work over which he
has had no control. In order to aid the finisher as
much as possible, I give a few hints herewith, for the
use of the workman whose duty it may be to prepare
and put up the work to be finished. The suggestions
given are equally applicable for hard and soft woods,
and I trust they will not be out of place.

First, then, see that the material is dry, free from
imperfections, of the full sizes required, and of such
variety of grain as may be suitable for the purpose
intended. Next, make all joints close and as near
perfect as possible, as on this point rests, in a great
measure, the artistic appearance of the work. Make
all angles sharp and clean, and all mitered mouldings
true and with perfect intersections. Never use glass
paper where a scraper can be applied, and when a
large quantity of mouldings of similar contour and size
are to be employed; it is always better to make a

scraper their reverse shape, and use it in cleaning and preparing the mouldings for the varnisher, than to use glass paper for the purpose.

The scraper illustrated by Fig. 51, and shown handled by Fig. 52, is an aid in producing surfaces more flat and regular than can be produced with the plane alone. Its use does not dispense with the plane; on the contrary, any surface on which the scraper

FIG. 51

is to be used must previously be planed as level and smooth as plane can make it. But the plane, in doing its work—no matter how sharp it may be, or how closely the back-iron is set up to the edge of the cutting iron, or how straight across the edge the cutting iron is sharpened—is liable to leave marks and ridges on the face of the work, which, on hard woods, are not effectively removed by the use of sandpaper alone. The scraper is used after the plane has finished its work, and previous to the final operation of sandpapering. In addition to the removal of the ridges left by the plane, the scraper is used for dressing up all kinds of cross-grained surfaces that occur in curly and figured solid and veneer woods; but as far as possible this use of the scraper should be avoided. Excessive dressing up of a cross-grained patch on a panel, a stile, or a table-top will most certainly show, and spoil the flatness and general appearance of the article when it is polished. As a rule, such excessive

FIG. 52

scraping is resorted to in consequence of the plane having been sharpened and set badly, or of some other technical oversight or unskillful manipulation.

The scraper is a thin and very hard steel plate, about 5 in. by 3 in., or 4½ in. by 2½ in., and slightly less than $\frac{1}{18}$ in. in thickness. The long edges are sharpened in a peculiar manner. Both of the long edges may be straight, as in Fig. 51, and at AB, Fig. 53, or one edge may have round corners of differing radii, as at

FIG. 53

CD. These corners are often useful in working up hollows and mouldings generally. The "straight" edge AB, it will be noticed, is not quite straight throughout its whole length. Near the ends the edge is gradually rounded off, to prevent the corners catching in the surface that is being operated on. In this respect, the commercial scraper (Fig. 51) is incorrectly shaped. The cutting power of a scraper depends upon, first, the quality and temper of the steel of which it is made; and, secondly, upon the proper formation of the burr or feather along its edges. Also, the faces of the steel plate must be perfectly bright and free from rust marks or indentations of any kind. It is by no means an uncommon thing to find any rusty piece of sheet steel—a piece of an old hand saw or try-square, for example—being used as a scraper. The smallest appreciation of why the scraper cuts would indicate how useless such material is for this particular purpose. On the other hand, scarcely anything better can be found for making a scraper than a piece of broken saw blade, provided the sides of it are still polished and bright. The fact that the saw was broken may easily be taken to indicate a higher temper than usual in the blade, thus fitting it exactly for the purpose of a scraper. The temper of an ordinary saw blade is not usually hard enough. Provided that there is enough elasticity to "give" in the operator's hands as it is being pushed along, the scraper should be nearly too hard for an ordinary file to touch. If it can be filed easily, then its edge will soon be gone. At the same time, if the steel is merely hard without the required amount of elasticity, the burr will strip off as it is applied to the work, leaving a coarse, jagged edge which is worse than useless.

The correctly sharpened scraper is a real cutting tool, and not, as its name suggests—and as in practice it often is—a mere abrader of the surface. When in good working trim, the scraper should, if desired, take off shaving after shaving perfectly uniform in thickness, and nearly as wide as the cutting edge is long. But such a performance is rarely required of it, and never when the plane has previously done its work properly. Too much emphasis cannot be laid on the fact that the proper duty of the scraper is not to make a surface, but to correct the irregularities on it. In explanation of the cutting action of the scraper, a diagrammatic figure is shown in Fig. 66. The figure correctly illustrates the cutting principle, though it does not represent a true section of the scraper. In use, the scraper is held firmly in both hands and tilted forwards, away from the operator, until the cutting edge grips the surface of the wood, exactly as shown in Fig. 54. It is then kept steadily at this angle, and made to cut a fine shaving at each stroke as it is being pushed away from the operator in the direction from E to F (Fig. 54).

The proper formation of the burr edge is of the greatest importance. Having procured a suitable

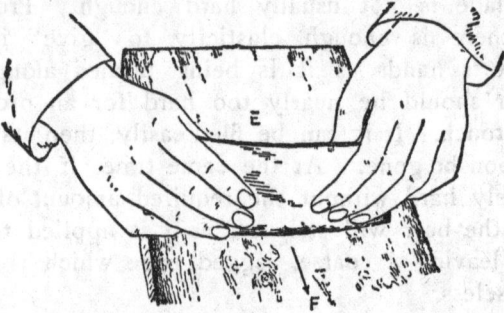

FIG. 54

steel plate, a usual but not recommended method is as follows: The plate is laid down on the bench as shown in Fig. 55. A narrow chisel, brad-awl or gouge, as shown in Fig. 56, is then laid on the scraper horizontally, and with considerable pressure is stroked backwards and forwards from end to end of the plate, G to H, Fig. 55. After about 10 or 12

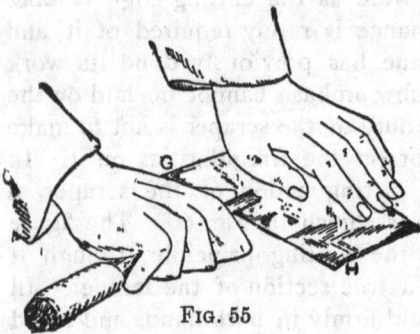

FIG. 55

strokes the scraper is turned over and the other side is treated in the same manner. This completes the first part of the process of sharpening. Next, the gouge is placed vertically against the edge, as shown in Fig. 57, and stroked to and fro with about the same pressure and firmness as the sides of plate received,

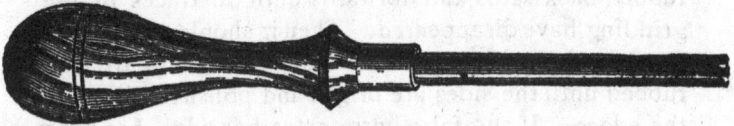

FIG. 56

and about the same number of times. Or else, for this second part, the scraper may be placed on its corner on the bench, as shown in Fig. 58, and held firmly in the left hand, while the right hand deftly strokes the convex side of the gouge with a quick action and considerable pressure, once or twice in an upward direction, as from I to J in Fig. 58. The scraper is then tried on the work to determine its sharpness. If, after this process, the scraper fails to cut satisfactorily, it is laid down on

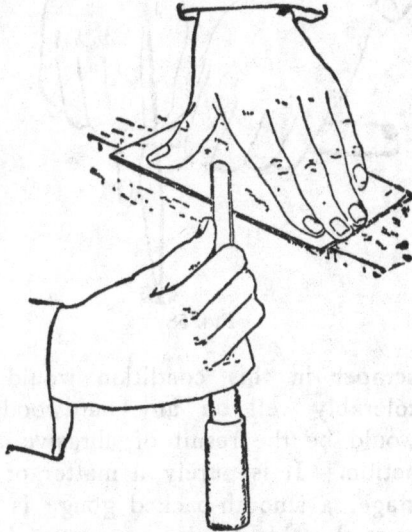

FIG. 57

the bench and the two processes are repeated. The following is a better method of sharpening a scraper: First, have the scraper ground so that its edge shall be straight and square, and slope up at the ends, as shown at A and B, Fig. 53. It should then be placed on its edge on a fine-cutting oilstone, as shown in Fig. 60, and rubbed backwards and forwards until all traces of the grinding have disappeared. Then it should be laid flat on its side, still on the oilstone, as in Fig. 61, and rubbed until the sides are bright and polished all along the edges. If any false burr or feather-edge has been created in this last process, the scraper must be set up on its edge again, as in Fig. 60, and rubbed a little more, until two perfectly square and sharp corners appear all along the scraper. Now, if due care has really been taken in making the corners perfect and square, the

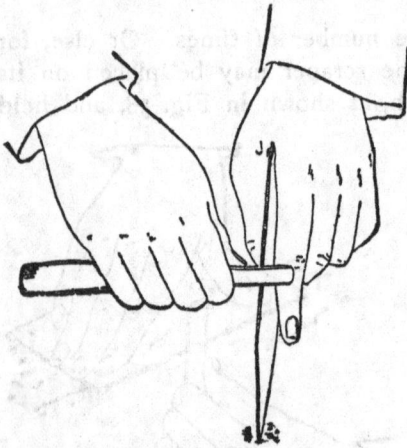

FIG. 58

scraper in this condition would produce shavings tolerably well on any hardwood; but the shaving would be the result of abrasive and not of cutting action. It is purely a matter of choice if, at this stage, a smooth-backed gouge is passed to and fro over the side of the scraper, as in Fig. 55, with the object of imparting a still higher polish to the plate of steel at the places where the burr is to be formed. But the greatest care must be taken not to press too heavily, and also to maintain a perfectly level position with the gouge, while the polishing is being done. As stated, this polishing of the side is not really necessary, and the scraper would cut very well without

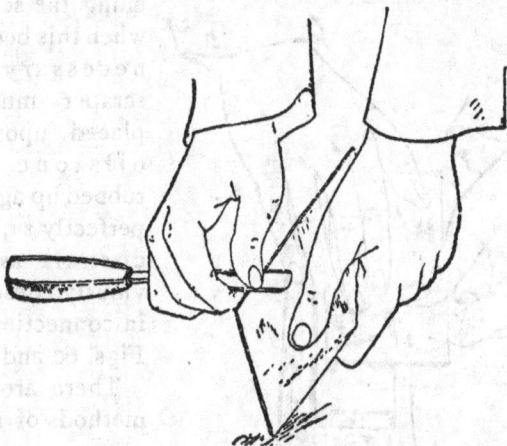

Fig. 59

its supposed assistance. Next, the scraper is placed on its corner on the bench, and a flat and smooth-backed gouge is passed once—or at most twice—along the edge. The scraper during this operation may be held either as already shown in Fig. 57, or as is here rec-ommended and shown in Fig. 61; or it may be fastened in the bench-vise. The aim in this

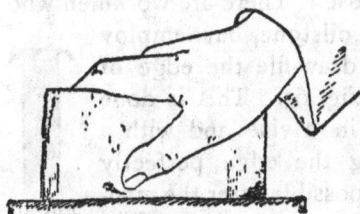

Fig. 60

particular action is the production of an exceedingly fine burr, scarcely enough to be called a burr at all, but a per-fectly regular bending over of the corner; the gouge must be very lightly pressed, or it will cause the burr to curl up too much, and also it must be kept nearly, though not quite, horizontal. With regard to resharpening the scraper, when this becomes necessary, the scraper must be placed upon the oilstone and rubbed up again to perfectly square corners as previously described in connection with Figs. 60 and 61.

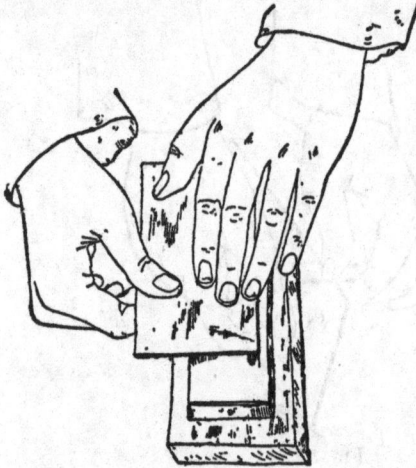

FIG. 61

There are other methods of sharpening scrapers which are made use of by some workmen; indeed, nearly every workman, after experience, will adopt such method as will seem to him to be the best for the purpose. There are workmen who seldom or never use an oilstone, but employ a fine file instead, and draw-file the edge of the scraper as shown in Fig. 63. This is done by placing the scraper in a vise and with a smooth, flat file making the edge perfectly square and as straight as possible after the manner of jointing a hand saw. Next place the file squarely across the edge and pass it from end to end of the scraper two or three times. This operation is known as draw-filing, a plan view of the position and direction of the file being shown in Fig. 63. Treat all four edges

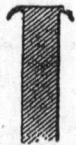

FIG. 62

the same way. The edge, as it will now appear, is
shown enlarged and exaggerated in Fig. 64. Remove
the scraper from the vise and lay it flat on the bench,
then taking the gouge, Fig. 56, rub out straight all the
wire edges, keeping the gouge on the scraper and passing
it quickly back and forth after the manner of stropping

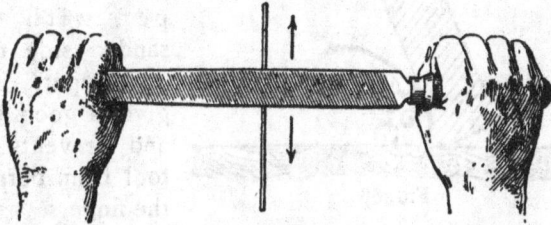

FIG. 63

a razor. The edge will then appear as in Fig. 65.
Now, taking the scraper in the left hand and holding
it firmly edgewise on the bench, place the gouge
across the edge, making a small bevel with the side of
the scraper, and draw it upward two or three times,
using considerable pressure. This will turn the edge

back as it was after filing, but it will
now be straight instead of grooved, and
smooth instead of ragged. All of the
eight edges must be treated in the same
manner, when they will appear as in
Fig. 66, and the scraper should now

FIG. 64 take off a shaving like a smooth-plane, FIG. 65
but much finer.

In order to resharpen the scraper it is not necessary
to go through the filing operation again for some time—
simply flatten out the edges and turn them again with
a little more bevel than before. This can be done

very rapidly. In order to avoid too many stops, always keep three or four scrapers at hand and sharpen them all at once. The best thing with which to hold the scraper is a piece of sandpaper, with the sanded side next the scraper. This gives a good grip and prevents the tool from burning the fingers.

Fig. 66

Some people file the edge of the scraper rounding, to prevent the corners from catching. This is not at all necessary, as the action of pushing the scraper bends it slightly, which raises the corners somewhat. The sharpening of a scraper, however, like the sharpening of a hand saw, takes considerable practice and no little knack; so if at first one does not succeed it is only necessary to keep pegging away at it until success crowns the efforts, for it is well worth all the trouble. It may be mentioned incidentally that a good burnisher may be bought all ready at

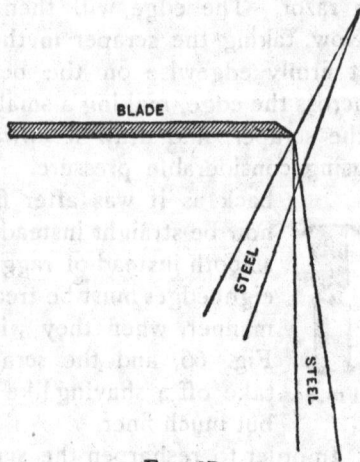

BLADE

STEEL

STEEL

Fig. 67

any good hardware store. Leather curriers use them for turning the edges of their knives and they are better than a gouge because of their being tempered harder and more highly finished.

The true theory of a scraper, for some purposes, may be described as follows: Where there is a lot of superfluous wood to remove, as in hardwood floors or other similar work, where it is not desirable to use a smooth plane, it is well to file or grind the scraper to a bevel, the same as a plane iron, and bring it to a keen edge on the oilstone; then proceed to burnish it. Hold the burnisher slightly at an angle with the bevel as indicated in Fig. 67, and draw it lightly across the blade. Then increase the angle and the pressure, repeating the process until the burnisher is at nearly right angles with the blade, after which run the burnisher back and forward a few times, first one side and then the other, as indicated in Fig. 68, when it is ready for use. When it becomes a little dull, do not turn the edge back, as many do, but use the burnisher, as shown in Fig. 68. A good blade will stand for two or three hours without filing or grinding.

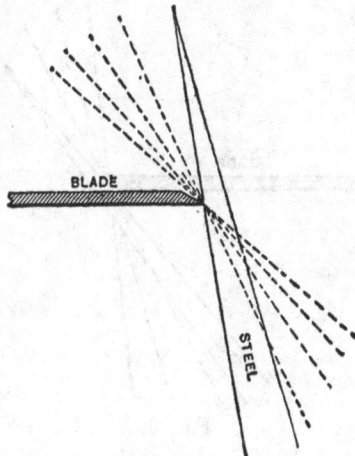

BLADE

STEEL

FIG. 68

For fine work file or grind the blade perfectly square on the edge and get it perfectly smooth on the oilstone. Then hold the burnisher as shown in Fig. 69, the dotted lines indicating how to start and the full lines how to finish. In all cases finish the operation as shown in Fig. 69. If unsuccessful the first time, do not give up the job, for the scraper is a tool that requires a great deal of practice in order to become expert in sharpening.

Defective Sharpening of Scrapers.—A frequent mistake is to put too great pressure on the gouge, and to stroke too much. One or two strokes should be quite sufficient to give the scraper the proper edge. At all events, after the scraper has had a couple of strokes of the burnisher or gouge, it should be tried, and if it does not grip the work properly, give it a few strokes more. Scrapers with a high temper require more burnishing than if soft or medium temper, but if overdone, the edge will crumble and it will not do clean work. Soft scrapers are easier handled than hard ones, but require a great deal more sharpening, and this sometimes leads to rough work, as the workman will not take time to repeatedly put his scraper in order. As before stated, the workman should have at least three or four scrapers on hand, and they should always be kept in good order. It is a good plan to have a little box or case in the tool chest purposely to hold scrapers, for two reasons: The workman will always

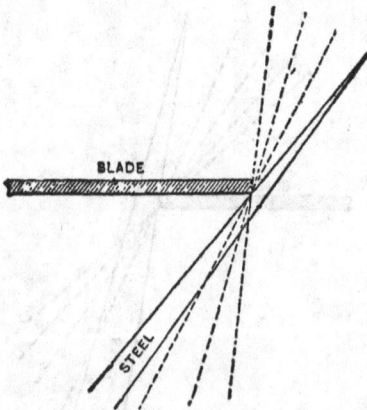

BLADE

STEEL

Fig. 69

know where to find them when wanted and, secondly, it will save the edges from being broken or damaged by being jolted or rubbed against other tools.

Varieties of Scrapers.—There is a new scraper in the market which is said to be superior to anything made in the scraper line. A representa-

FIG. 70

tion of it is shown in Fig. 70, which is taken from an advertisement. I do not know of my own knowledge how the scraper works, but, judg-ing from its shape and setting, I should think it theoretically cor-rect.

FIG. 71

Other shapes of scrapers are shown in Figs. 71 and 72. The first of these is intended for smoothing up hollows and rounds. A number of these should be kept on hand, with curves of various radii. Fig. 72 shows a scraper made purposely for clean-ing out hollows of various curvatures, and will be found very useful on work

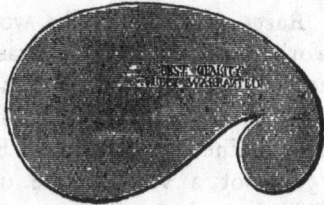

FIG. 72

where there are a number of curves or other hollow mouldings. Often the workman may have to make scrapers to suit certain kinds of work, and when such is the case, he should make it a fixed rule never to change them, but to keep them in his tool chest, and make others, when necessary, from new material. An old hand saw blade generally answers quite well for scrapers, and one saw will make a couple of dozen scrapers of different sizes and shapes.

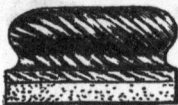

FIG. 73

How to Use Sandpaper.—For properly using sandpaper a rubber is needed. A piece of mahogany or clean pine, 5¾ in. by 3 in. by 1 in., shaped as in Fig. 73, answers well if a piece of sheet cork is glued on the face as shown. Fold a piece of sandpaper, 6 in. wide and about 10 in. long, into three, place it sand side downwards, and put the face of the rubber on the middle division. Grasp the rubber so that the ends of the sandpaper are held firmly on its back and sides (see Fig. 74), and work then can be commenced.

A solid rubber about one inch thick makes an excellent block for the purpose. A piece of rubber belting glued to a piece of basswood also makes a good rubber block.

Rasps and Files.—The woodworker occasionally uses rasps, and these generally are halfround, though sometimes flat. The cabinet rasp shown in Fig. 75 is not a very coarse one.

FIG. 74

Cabinet and wood rasps range from 4 in. to 14 in. in length, and at 12 in. and less the price is about 4 cents per inch. The extra 2 in. in a 14-inch rasp adds nearly 50 per cent to the price. Files also

are used, for smoothing up hardwood. When a file is used, it should be pushed and drawn in the direction of the grain. The file usually employed by carriage makers for smoothing up spokes and other work is generally a half-round bastard file, and when properly used makes exceedingly smooth work. A finer file than the bastard is shown in Fig. 76. This is sometimes used for finishing narrow flat work.

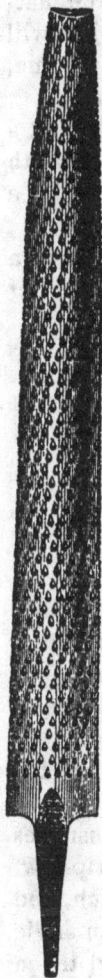

SAWS FOR WORKING HARDWOOD

Saw-filing is an art unto itself, and few men ever attain the art to perfection. To file a saw in such manner that the working of it is a pleasure requires an amount of skill and a high order of technical knowledge that can only be acquired by thought and experience; yet a properly set and filed saw is a real necessity in the making of good joiners' work, and, while every workman cannot become an expert saw-filer, he ought at least to possess such knowledge of the art as will enable him to set and file his own saws in such a manner as will enable him to execute such work as he may be called upon to make; and to this end, I propose to offer

FIG. 75   FIG. 76

a few instructions and remarks that will assist him in managing his own saws without the employment of expert aid. It is not my intention to go into the matter of saw-filing to any great extent, as that subject will probably be discussed at length in another volume, but what I do offer will, I trust, be up to the mark.

All workmen in wood require two saws, namely, a cross-cut saw, and a rip-saw. The shape of the teeth in these saws differs, also the size, and each requires a special form of tooth and a different style of filing.

Many workmen think that so long as the tooth of a saw has a sharp edge the shape of the tooth is a matter

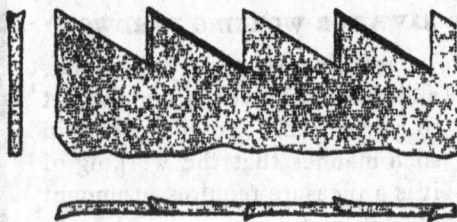

FIG. 77

of small importance, and as a result of this ignorance or indifference they are always in trouble with their saws, and their work becomes much more laborious than if proper attention had been paid to the shape of the teeth. Substances of different texture cannot be cut advantageously with the same tool; in fact, the tool must be adapted to the work if the best all-round results are to be produced. Fig. 77 illustrates a form of tooth suitable for an ordinary hand rip-saw. The tooth points number about four to the inch, and the front of the tooth is upright, that is, at an angle of 90°. The face of the tooth should be filed to an angle of 85° to 87°, or 3° to 5° from the square. Some experts contend that the teeth of a rip-saw should be filed dead square. The object in filing them a little on the bevel is that the teeth may cut more freely and easily when they have become a bit dull, there being then what is termed a little clearance cut in the teeth. In nearly all timber there is, it is well known, a certain

FIG. 78

amount of fiber to be cut either directly or obliquely across, and teeth that are filed square will not, whether they are sharp or dull, divide this fiber so easily as teeth that have a slight bevel.

The illustration, Fig. 77, shows a saw tooth that works quite well in the softer woods, but for the harder

FIG. 79

woods a little more plane or angle on the cutting edge would cause the saw to cut with less labor; and if they were about five teeth to the inch, it would cut with ease any kind of hardwood.

The teeth shown in Fig. 78 are well suited to a hand saw used for cross-cutting soft wood. The tooth points in this saw may number five or six to the inch. The front of the tooth slopes at an angle of about 105°. The face of the tooth in sharpening should be filed to an angle of from 55° to 60°. The softer the timber that is to be cross-cut, the more acute should be the angle of the teeth, as the keener edge separates the fibers more easily. Fig. 79 shows a form of hand saw

FIG 80

tooth suitable for cross-cutting hardwood. The number of tooth points may be from six to eight to the inch, and the front of the tooth should slope at an angle of 110° to 115°, according to the hardness of the timber to be sawed. The face of this tooth should be filed to an angle of 70° to 75°, because the cutting edge must

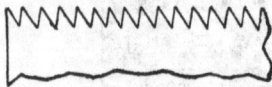

FIG. 81

be less acute owing to hardwood fibers being more compact than those of soft wood.

The whole number of saws made use of by the woodworker amounts to some six or eight, comprising the rip, cross-cut, hand, panel, tenon, dovetail, bow or turning, and keyhole. The hand saw type includes the hand saw proper, the ripping, half-ripping, and panel saws, all of similar outline, but differing in dimensions, and in form and size of teeth. There is no sharp distinction between these tools, as they merge one into the other; yet at the extremes it would be impossible to substitute the ripping and panel saws one for the other. The hand saw, however, which is a kind of compromise between extremes, is used indiscriminately for all purposes.

A tenon, or back-saw, is shown in Fig. 80. It is made in different lengths, and the blades are not all made the same width. The dovetail saw is a back-saw with a very thin blade; it is not much used in this country, only by Europeans. It is intended especially for very fine work.

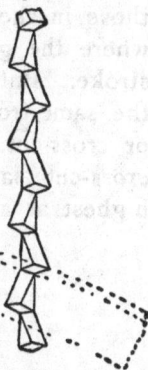

FIG. 82

The other saws named are for special purposes, and need not be described here.

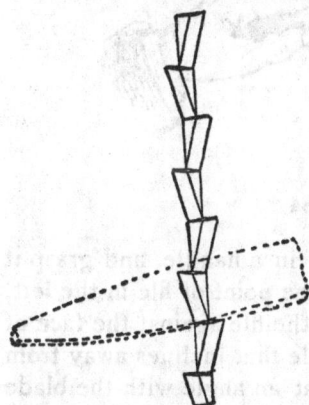

FIG. 83

The methods of filing, however, are about the same as described for the saws first named, with the exception of the bow or scroll saw. This saw, which is intended to cut with and across the grain, or obliquely to it, should have its teeth filed with a sharper plane than a rip-saw, but not quite so sharp as a cross-cut saw; in other words, it wants a hybrid tooth, "between and betwixt" a cross-cut and a rip-saw tooth. (See Fig. 81.)

The set of a saw is important, and this is shown in

Figs. 77, 78, 82 and 83. The teeth at both point and butt of a hand saw should be very slightly smaller than those in the middle, as it is at the last-named point where the greatest force is exerted in every down stroke. But it is absolutely necessary that the set is the same from point to butt of every saw, whether rip or cross-cut. The middle of the cutting edge of a cross-cut saw should be slightly rounded, being highest at about the middle. The saw being still in

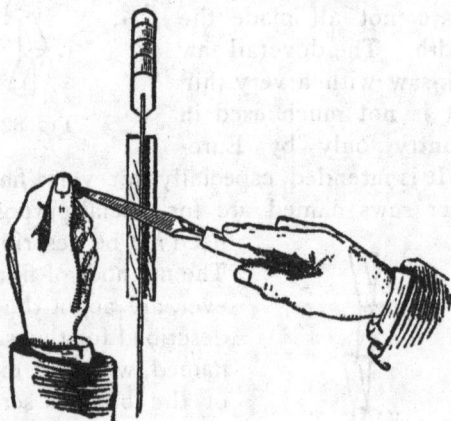

FIG. 84

the saw vise, insert the file in a handle, and grasp it with the right hand, taking the point of file in the left, as shown in Fig. 84. Place the file against the face of that tooth nearest the handle that inclines away from the worker, holding the file at an angle with the blade or saw as shown in Fig. 84. Then lower the right hand to about the angle shown in Fig. 85 (which shows the left hand removed). The file should be held obliquely across the saw blade, as in Figs. 83 and 84, the point end of the file being inclined towards the saw handle as illustrated by Fig. 84. Gently push the file forwards, lifting it at the end of stroke, returning it, and again pushing it, until the point of the tooth has a keen edge. Repeat this upon each alternate tooth until all upon one side are sharpened. Serve the other side in the same manner. Be careful not to press the file against the back of the tooth, as unevenness will surely result.

Keep all saws slightly rounding on the edge. The rip-saw and the hand saw may have a full quarter of an inch rounding with advantage. All saws, except circulars, have a constant tendency to get hollow, and this must be prevented; and the only way to prevent it is to file

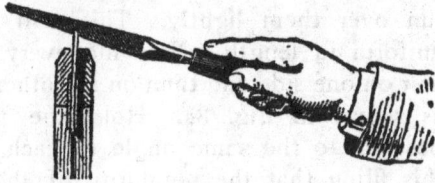

FIG. 85

the teeth down by passing a partly-worn file along the edges till it touches every tooth. Then, in filing the teeth, take care only just to take out this bright mark—not one touch more or that tooth will be shorter than its neighbors. A saw properly sharpened, and in good order, has each tooth do its proper share of cutting and no more.

Sharpening Tenon Saws.—Sharpening tenon saws is practically the same as sharpening hand saws. It may happen, however, that some of the teeth will be much larger than others, this being due to the file not having been held at the same angle in sharpening each tooth. Fig. 81 shows the saw teeth improperly sharpened, the front of the large teeth inclining much

more than the front of the small teeth. If the teeth
in one range are found to be filed smaller than those
in the other, file the back of the smaller teeth to a
more acute angle, keeping the file at the same time
well against the front of the other tooth at the bottom, and see before the filing is finished, that the front
of the next tooth has been filed up to its point, as it
is the front, not the back, of the tooth that does the
cutting. To regulate the teeth of the saw, file every
tooth in succession, shooting the file straight across
the teeth. After filing all the teeth from one side,
turn the saw, and file as before from the other side.
When the teeth are fairly regular, the flat file may be
run over them lightly. This will bring the teeth
uniform in length. Now file every alternate tooth,
first on one side and then on the other side of the saw,
as shown in Fig. 84. Hold the file as nearly as
possible to the same angle in each case, as it is on
this filing that the regularity of the teeth depends.
When all the flat places caused by the file when
topping the teeth disappear, cease filing, as any further
filing may cause low teeth, which tend to make the
saw cut irregularly. When the saw is so filed that all
the teeth get their equal share of cutting, the saw may
be said to be well filed. When the teeth are filed as
shown in Fig. 83, there is a larger and better cutting
edge than with the one more obtuse. If the points
only of the teeth are allowed to do the work, the action
is a scratching and not a cutting one.

One of the great difficulties in hand sharpening is to
get the bevels of the teeth exactly alike. A number
of mechanical arrangements to guide the saw and
effect this object have been tried with more or less
success. In one of the best of these devices a circular

casting is divided and indexed from its center each way, giving bevels for each side of the saw, or square across. The file is fitted into a guide, and is held by a set-screw. The index shows the pitch at which the file is set, and a rod passes through holes in a graduating ring and guides the file. The frame upon which the ring is held slides in grooves cut on each side of the vise in which the saw is fixed; a table connected with the guide is arranged and indexed so as to give the required bevel and pitch for the kind of saw to be filed, and it is only necessary to set the ring for the bevel, and the indicator for the pitch, and the apparatus is ready for use. As the filing is proceeded with from tooth to tooth, the frame follows, giving to each tooth on one side of the saw the same bevel, pitch, and size as on the other, thus leaving the saw, when filing is finished, with the teeth all uniform in size, pitch, and bevel, so that each tooth will do its share of cutting equally

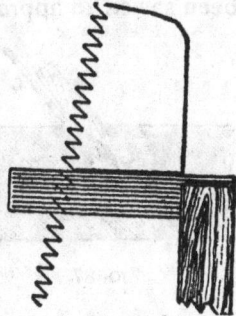

FIG. 86

with the others, thus turning out more and better quality of work with less expenditure of energy. An old-fashioned way of getting the right angle of a tooth of a hand saw in filing is shown by Fig. 86. A hand saw blade is narrowest in width at the point, and broadest at the butt; and the slope of the back, compared with the line of teeth, is almost always uniform for all saws; and if a square be placed against this back, a tooth may be filed whose cutting edge is perfectly in line with the edge of the square. All the teeth being thus filed and afterwards set, a saw which will answer general purposes is produced, and one which will suit the worker who has but one hand saw. It will cross-cut soft woods and rip hard woods, thus being a kind of half-rip. Moreover, this square may be used as a gauge, the teeth not necessarily being filed as shown; and if the rake or lead be very much, an adjustable bevel may be used.

PLANES GENERALLY

In workshop practice, planes are the tools chiefly used for smoothing the surface of wood after it has been sawed to approximate size. In its simplest form,

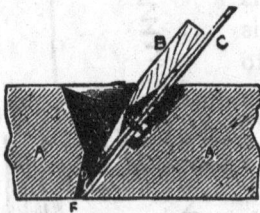

Fig. 87

a plane is a chisel firmly fixed into a block of wood by which it is guided in its cut, and the amount of wood taken off in the form of a shaving is regulated to a nicety. In fact, such a simple tool actually is used sometimes, when a proper plane of the requisite shape and of a suitable size cannot be procured. To make the construction of an ordinary plane quite clear, a section of one is illustrated by Fig. 87, in which A shows the section of stock; B, the wedge; C, cutting iron; D, back iron; E, the screw for fastening irons together; and F is the mouth through which the shavings pass upwards. A plane is simply a copying tool, and a notch in the plane-iron at once proves that the pattern produced corresponds with the edge of the plane-iron, and all the imperfections of this edge will be copied on to the stuff. In all planing operations the edge of the tool is the pattern, which is copied in reverse on the wood. If a hollow is required to be produced on the wood, a tool is used with a round edge of exactly similar form to the hollow we wish produced. In machine planing the pattern is the edge of the tool, which produces a similar surface upon the wood.

Direction of Grain in Planes.—For flat planes such as jack, try, and smoothing planes, the grain must be straight and, of course, run lengthways of the tool. The wood is selected from a center plank of beech as near to the bark as possible; in all planes, the wood nearest the bark, that being the hardest part of the wood, becomes the sole or working surface. The wood is seasoned thoroughly, and never is used until at least three years after cutting. Moulding planes mostly work on the spring, and need not have the grain so straight as flat planes. Hollows, rounds, and rebate planes are exceptional, however, and should have the grain as straight as possible, because the rebate plane is cut right through, and is liable to cast or warp if not quite straight-grained; and most of the hollows and rounds, being thin, are liable to cast also if not of straight, mild, and well-seasoned wood. Wood for plane making should be as free from knots as possible.

Jack and Trying Planes.—The jack plane, Fig. 88, is the first applied to the wood after it has been sawed. This plane is always employed to remove the roughness of the work before finishing up with trying and smoothing planes. It is made up of five parts— namely, the stock (which should be 17 in. in length), the toat or handle, the wedge, the cutting iron or cutter (2⅜ in. broad), and back iron. Immediately behind the iron is a handle, which, in use, is grasped only by the right hand in planing fir; but in heavy planing, and especially in hardwood, it is necessary to place the left hand across the front of the plane to press it down, to cause the iron to take hold of the wood. When using both hands to the plane, the left is placed with the four fingers lying across the top near the fore end, the thumb passing down the near side. Well-seasoned beech is a suitable wood for the stock. The construction of the trying or truing plane is almost exactly the same as that of the jack

Fig. 88

plane, but it is much longer, so as to produce truer surfaces.

Using the Plane.—When using the plane, always keep the plane well oiled in front of the iron; be sure and work with the grain as shown in Fig. 89, whenever possible. A far thicker shaving can be taken off smoothly in this way than if worked as shown in Fig. 90, which is working against the grain and requires more careful work with a more finely set plane. Do not use the plane at an uncomfortable height. When the board to be planed is in position, and the worker takes hold of the plane ready to begin work, a line drawn through his elbow and wrist should be rather lower than higher at the wrist, though if the forearm is

level it will do fairly well. Do not attempt to take off
thick shavings at the outset, and do not be dis-
appointed if a shaving is not taken off from end to
end at the first trial. If the wood has any hollow in
it, it will be impossible to do this, and even if it is
perfectly straight it is sometimes difficult to do it.

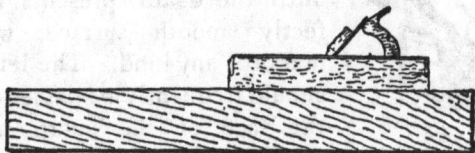

FIG. 89

There is always a tendency to plane off too much near
the bench stop, as the arm is apt to be raised at this
point. Errors of this kind can be avoided by careful
practice and attention. It is a good idea to try to
plane the center of the material rather than the margin,
for if a good plane is in proper order it is impossible
to make the work too hollow or concave; whereas,

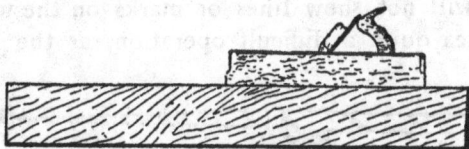

FIG. 90

however good the plane, careless use of it can and will
make the work convex in every direction.

In making use of the smoothing plane for finishing
up stuff, it should be held as shown in Fig. 91, the
right hand grasping it firmly at the back, whilst the
left hand steadies it in front. In starting, the tool is

applied to the fore-end of the board, and gradually worked backwards, thus taking out any marks previously used tools may have made. It should be held firmly, and lifted sharply at the end of stroke, or a mark will be left where the plane finished. The work is continued until the board presents a perfectly smooth surface, without marks of any kind. The left hand should frequently be passed across the face of the board, as any marks made by the plane can be readily detected in this manner. It may also with advantage be held so that the light passes across its surface from the side, thus showing up imperfect planing. Should there be too much iron out, a few blows with the hammer at the back of the plane will draw it back. Take care to tighten up the wedge again. A few drops of linseed oil applied to the face of the plane will facilitate the planing.

FIG. 91

The proper method of setting an iron in a plane so that it will not show lines or marks on the work, is sometimes quite a difficult operation, as the iron is

FIG. 92

expected to be perfectly square and straight on the face. It is always well to either round off the corners of the iron slightly, or grind them with a slight bevel, which will keep the corners from "digging in."

The smaller, or break-iron, whose office is to bend up the shaving somewhat sharply so as to ensure the cutting of the other iron, and to prevent its splitting off the surface of the work, should be placed so as to come within one-eighth of an inch of the extreme edge of the cutter for rough work, and within one-twentieth for finer or finishing work. The two should then be placed in position so that the edge projects the smallest possible degree below the sole. The position can only be determined by looking carefully along the

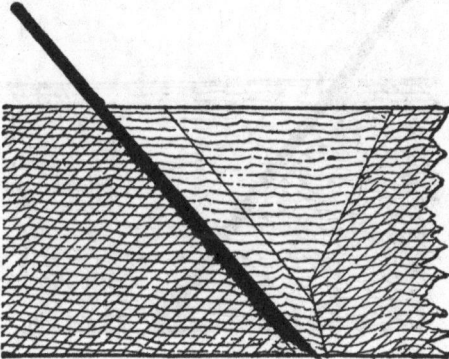

FIG. 93

bottom of the plane, with the front of the same next to the eye, as in Fig. 92. The edge will, if correctly formed and placed, appear quite parallel with the sole. It is then ready for use. The same rule applies to the small as to the large planes, except that in the jack-plane the iron projects rather more, as it is used for roughing down a surface. The trying-plane, which is longer, as before stated, and intended for edging boards which are to be joined lengthwise, is always very finely set, and the mouth is narrow. The break-iron is also set very close down to the cutting edge. The longer the plane the more accurately level and true will be the work done by it.

New planes are often a source of trouble owing to the shavings getting fast in the mouth, the plane refusing to take any more until the wedge and iron have been removed, and all cleaned out. The cause of this difficulty is that the mouth of the plane is too narrow, as shown in Fig. 93. Sometimes workmen cut

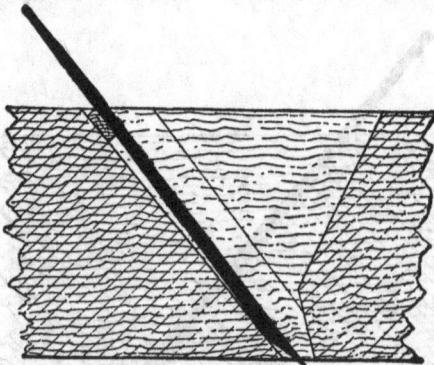

FIG. 94

a little out with a chisel, but in many instances this results in spoiling the tool, because in a short time the bottom of the plane wears away, and the mouth gets larger, subsequently getting so large as to require a piece to be set in to lessen the opening. The mouth of the plane can be kept open as long as needed by gluing a strip of soft leather, about three-fourths of an inch wide, in the mouth of the plane under the top end of the iron as shown in Fig. 94. When the plane bottom is sufficiently worn the leather can be removed and the iron put back into its original position. The leather causes the iron to be more upright, and consequently there is a larger opening in the mouth (two thicknesses can be put under if necessary).

Pitch of Plane Cutting Iron.—To assist the reader to understand correctly the principles of plane-iron sharpening, the following information is given. The seat of the plane-iron is made at different angles, to give the pitch to suit different kinds of work. The four angles most in use are as follows: Common pitch, in which the seat for the back of the iron is at an angle of 45° from the sole (this inclination is usually employed for all planes for soft wood); New York pitch, which has an angle of 50°, and is adapted for use with mahogany and other hard, stringy woods; middle pitch or 55°, and half pitch or 60°, which are employed with m o u l d i n g planes, the former being for

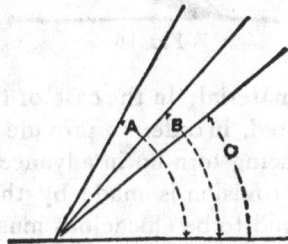

FIG. 95

soft wood and the latter for the harder kinds. Fig. 95 affords an idea of three angles, A giving the set of a half-pitched plane, B that of an ordinary plane, while C shows the inclination of an extra-pitched plane. The pitch or angle at which the cutter is set is of importance. There are three angles involved in this case: (1) The angle between the cutter and the surface of the work; this should be as small an angle as possible. It is obvious that if the surfaces of the cutter and the work were perfectly parallel, the cutter would glide over the surface without cutting, except under great pressure. By making the cutter edge rather than its whole surface touch the work, the tendency to cut and to continue the contact is secured. The angle, which may be called the clearance angle, or the back angle, should only be enough to secure this condition of contact. (2) The angle of the cutter itself. The more acute this angle the better, if only the material will stand the strain and face the work without losing its edge. (3) The front or remaining angle may be found by subtracting both (1) and (2) from 180°, if dealing with plane surfaces, and is available for the passing away of the waste material; in the case of the plane, however, this is limited, in order to provide means to prevent the shaving being torn up in advance of the cutting action. This provision is made by the front portion of the plane, and to be efficacious must be in contact with the work and as near the edge of the cutter as possible to allow waste to escape. A few experiments with a knife will show that for soft materials a slight angle is best. This involves a thin knife, and its side almost in contact with the material to be cut; but as harder things are tried the stiffness of the cutter, and the consequent angle, must be increased, not because it is merely desirable, but because it is absolutely necessary to have a stronger cutter. The more upright an iron is set, the less

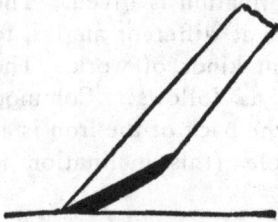

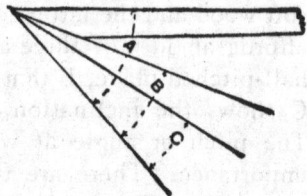

FIG. 96

FIG. 97

liable is it to tear up the wood in planing; but in the same degree the iron loses its edge more quickly, is more likely to jump and chatter, and is more laborious to work. In toothing planes the irons are set upright, and in "old woman's tooth," planes or routers nearly upright. In working with an upright iron, the action of the tool is a scraping one, while the more the iron is inclined the nearer it approaches the action of splitting the wood. For this reason an ordinary or extra-pitched plane is more liable to splinter up the surface of a piece of work; but this accident may be in great measure prevented by using a properly adjusted back iron. Moulding planes, rounds and hollows, bead planes, and others that work without a back iron, are usually preferred set to the half-pitch angle; while for

FIG. 98

working on end-grained stuff, extra-pitched tools, such as shoulder and bull-nosed planes, give the best results. In examining the side of a plane-iron, it is found to be made up of an iron back faced with steel. The steel, welded on to the iron and distinguished by its brighter color and finer grain, acts as a cutting edge, the iron being required to give sufficient stiffness to prevent chattering. When newly ground and sharpened, a plane-iron has three angles, one due to the pitch, A, Fig. 97, one made by the grindstone, B, and one made by the oilstone, C. The angles A and B do not alter, but C gradually becomes more acute with sharpening on the oilstone, until it lines with the face of the plane, as in Fig. 98, when the iron refuses to cut properly, and requires grinding. The pitch angle A (Fig. 96) varies in planes by different makers, as described above.

Oilstones for Sharpening Plane Irons.—A good oilstone, capable of putting a keen edge on the plane irons, is a necessity to all woodworkers, but more particularly to the hardwood finisher. The Washita stone, manufactured by the Pike Mfg. Co., or other fast cutting stones are useful for removing the waste or superfluous metal from the iron when it is too thick; but it is seldom that one of these stones can be trusted to leave a keen edge on the tool. Turkey and Washita stones are the only ones that have the two good qualities combined of cutting quickly and leaving a good edge.

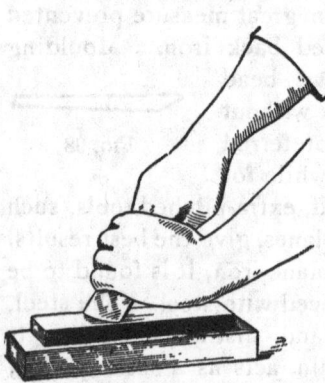

FIG. 99

Most other oilstones are slow cutters, but they are to be relied on for leaving a good, keen edge. It is necessary that the oilstone should be kept perfectly level, or it will not be possible to get a true edge. The stone should also be free from grit, or the iron will be gapped in sharpening, and will leave ridges on the planed work. In sharpening the iron after it has been newly ground, the hands should be kept low to make the bevel correspond nearly with that made by the grindstone. As time goes on when the iron is resharpened the hands are kept a little higher upon each occasion (see Fig. 97), until is becomes thick, as in Fig. 98; then it must be again put upon the grindstone. Some workers find it convenient to use two oilstones—one as a quick cutter, to some extent superseding the grindstone, the other for finishing the edge.

In using the oilstone, first put a few drops of good oil
upon the stone, and grasp the iron as in Fig. 99; the
right hand is at the top, and the thumb and fourth
finger pass under. Place
the whole of the fingers
of the left hand upon
the iron, with the thumb
at the back, as seen.
Now put the cutting
edge (previously ground
to a bevel) upon the
stone in an oblique di-
rection, as shown in

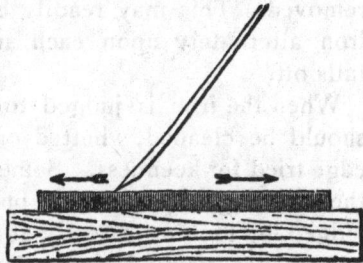

FIG. 100

Fig. 99, bearing in mind the previous remarks on the
necessary inclination. The iron should now be rubbed
up and down the stone, pressing it down with both hands.
If the edge had been examined before placing it upon
the stone, it would have been found to show a fine white
line. The object of sharp-
ening is to remove this,
which must be done by
rubbing on an oilstone.
Having accomplished this,
turn the iron face down
upon the stone, and rub it
lightly a few times (see
Fig. 100). The iron should
now have the appearances

FIG. 102

FIG. 101

indicated by Fig. 101. If the face of the iron has not
been kept perfectly flat, it would appear as Fig. 98,

and would be of no use as a cutting iron. If the iron has been rubbed too long, a wire edge will appear and utterly spoil the cutting properties of the iron unless removed. This may readily be done by rubbing the iron alternately upon each side until the wire edge falls off.

When the iron is judged to be sufficiently sharp, it should be cleaned, whetted on the left hand, and its edge tried for keenness. Some try the edge by passing the thumb gently across it, but its sharpness may be judged by looking directly at it. In a sharp tool the edge is not visible to the naked eye, while if the iron should be blunt, the edge will be seen as a bright line.

SECRET OR BLIND NAILING

Secret nailing is sometimes called "blind nailing," also "chip nailing" and "sliver nailing," and is the art of finishing work in a manner which leaves no mark of nail holes or screw heads exposed to view, which, under the ordinary method of doing work, would require puttying before the painter could apply the finish.

The process of secret nailing is only used on occasional jobs of hardwood finish where an extra fine job is required. The process is very simple and can be followed by any mechanic of ordinary skill and ability. Take a very thin and sharp paring chisel, ⅜ to ½ inch wide, to raise the "chip." A sharp knife should be used to make two cuts with the grain of the wood, the width of the chisel apart, to keep the sides of the chip from splitting. The chisel should be set at a steep angle at first, till the proper depth is reached, and then made to turn out a piece of wood of even thickness, about a sixteenth of an inch or near it, and of sufficient length to admit of driving the nail or screw. Care should be taken in raising the "chip" not to give it too sharp a curve or too great a thickness, as it is liable to break off while being straightened out again.

Some mechanics prefer a gouge for raising the "chip"; the gouge should be ⅝ to ¾ of an inch wide and of a quick curve. In this case no knife is needed, as the corners of the gouge will cut the wood as it advances. The cut being made and the "chip" properly raised, a nail or screw may be driven in.

See that the nail or screw head is sunk below the surface of the recess, so that the "chip" will fit back in again without any obstruction. Now take properly prepared glue

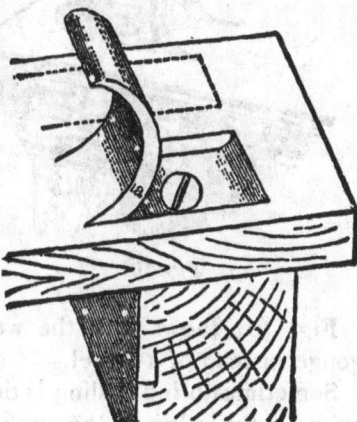

Fig. 103

and apply to the "chip" and recess, and press the "chip" firmly in place, rubbing the face with a smooth block till the glue holds, and finish by using a little sandpaper.

If there is any difficulty about the "chip" breaking off, moisten the wood with a little water applied with a sponge to the part where the "chip" is to be raised. This will be found to be a great advantage if working on brittle wood.

As stated previously, when putting up hardwood

finish, where it is desired that no nail or screw heads should be in sight, it is always better to make use of glue whenever possible; this will be found to be much more satisfactory than blind nailing.

The methods of secret nailing, as described in the foregoing, are shown in the diagrams, Figs. 103 and 104. The first shows how a square chip is raised so that a screw or nail can be put in place, after which the raised chip *a* can be glued down in place, covering the head of nail or screw.

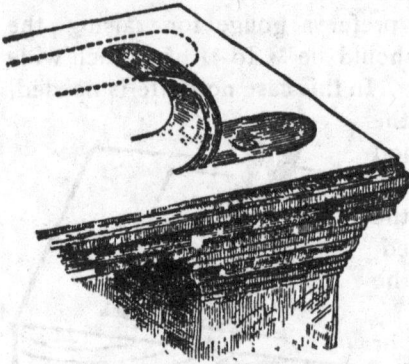

FIG. 104

Fig. 104 shows how the work is done by using a gouge instead of a chisel.

Sometimes blind nailing is done by driving headless nails in the edge of the work, "toe-nailing" them in the work just as matched flooring is laid. This is not a satisfactory way to do work and is not recommended, though there are cases where it will answer quite well.

PART TWO

WOOD FILLERS AND HOW TO APPLY THEM

There is no part of the art of wood-finishing that is more important than that of the filling, and the greatest of care should be exercised both in the choice of filler and the manner in which it is applied. The stain given to it must also be considered, in order that the color and texture of the wood being finished may not be disfigured or spoiled.

Fillers are used by all expert polishers for much the same reason that size is used before varnishing—viz., to prevent immoderate absorption of the polish by the wood. Polish, or even thick varnish, when applied to wood, sinks in or is absorbed in places, instead of remaining on the surface in a uniform coat. Here and there it will be observed that the polish or varnish has given more gloss than elsewhere. Where the gloss is brightest the varnish has sunk least.

The grain may be filled up by going over the wood with polish till the pores are closed, and some beginners may want to know why anything else in the nature of a filler should be used. The reason is that comparatively valuable French polish need not be used when a cheaper material serves the purpose, the use of which also saves time. Woods that are open in the grain and porous specially need a filler, while fine, close-grained woods do not, and may be polished without. Still, a suitable filler can do no harm to any kind of wood, however fine the grain may be, so there can be no disadvantage in going over it with one preparatory to polishing. Though it may be a slight waste of time, a preliminary rub over with polish suffices when working on a fine wood, such as olive or rosewood, which are both close and hard. To attain the desired thin, glossy film of shellac, which shall not be liable to grow dull unreasonably soon, the woods ordinarily used in furniture—ash, oak, mahogany, walnut, cherry, etc.—should have the grain filled, for they are all of comparatively open grain; ash and oak, being especially coarse, are called by polishers "hungry woods." Polishers usually give such woods one or more coats of spirit varnish as an aid to filling up the grain.

Before commencing the process of filling-in, thoroughly brush all dust out of the grain of the wood, for this is wood-dust, sand from the sandpaper, and dirt—all inimical to grain luster if mixed up with the grain stopper.

Amongst the best "fillers" is a preparation manufactured in Bridgeport, Conn., and known as "Wheeler's Wood-filler," and though it may cost a little more than home-made or other fillers, it is certain to give satisfaction. This filler has for its base a form of mineral silica in an atomic shape, which permits it to be ground or pulverized into a very fine, dust-like condition, in which each particle assumes a needle-pointed form, which enables them to enter into the pores of the wood and give to the work a gloss-like surface.

For filling a cheap class of work, many polishers content themselves with giving the work one or two coats of glue or patent size, heavily stained by the addition of some dry pigment. For mahogany finish add Venetian red till it gives quite a red tinge; for walnut add brown umber; for pine, add yellow ocher.

Apply the size hot with a brush, and rub it in lightly with a piece of rag, finishing the way of the grain, and taking care in the case of turned or moulded work to get the filler well in the recessed parts. Of course, work that has been sized will not need filling-in.

As many different kinds of fillers are used, and each has its advocates, it will be advisable to name the principal fillers used in the trade, and to make a few remarks about each, so that learners can experiment with them, and perhaps finally fix on that which may seem to suit best. All will be found reliable, for good work is turned out by polishers with any of them, and even an extremely prejudiced individual would hesitate to say that any one is really bad, though he uses only that which suits him best. Sometimes, owing to the price, he uses the easiest and quickest, irrespective of its quality.

Wood-fillers ready for use are made for most kinds of wood, and, as a rule, they require only thinning with a little turpentine. When it is desired to make a filler instead of purchasing one ready made, proceed as follows: Take a portion of either china clay or corn-flour; add boiling linseed oil, and stir until the mixture is of the consistency of putty. Then add patent dryers and thin with turpentine. If the wood on which the filler is to be used is to be kept light in color, use raw oil and the lightest variety of dryer. Further remarks on home-made fillers will be found later on.

In woods employed for house and cabinet work there are two distinct natures; therefore different treatment is required in finishing. First, there is the coarse or open-grained wood, having its surface perforated with innumerable pores or cells. In order to obtain a

smooth and even finish, these pores should be filled
up to a level with the hard grain, or, as the grainer
would term them, the "lights" of the wood. Next
we have the fine or close-grained wood, which, like
the preceding, also contains these pores, but they are
of a very fine character, and simply sealing them up
with a liquid filler will enable one to produce a fine,
smooth finish, which we shall consider farther on.

I will now return to the open or wide-grained wood,
which requires more attention and care than the
closer-grained wood, if the same results are expected.
There are many things that will serve the purpose of
fillers and make pretty fair work; among these may
be mentioned china clay, silver white and corn-starch;
the best of which is probably the last named. This is
well adapted to the work, being equally useful with
light, or when colored with dark wood; one fault with
it is that it never hardens. China clay—the English
is the best—makes an exceedingly good filler; it is
light in color, very fine and dries as hard as cement.

There are a hundred ways of preparing fillers for
use, and nearly as many different materials for making
them; we would recommend, however, that wherever
it is possible, Wheeler's patent wood filler be used;
though we are aware that in many cases it may not
be advisable to use it, and to meet these rare conditions
the following mixtures may be substituted.

A filler should be so mixed that the greater portion
of the vehicle will penetrate into the wood, leaving
the pigment on the surface to be rubbed into the
pores, and still retain enough combining property to
form a hard and impenetrable surface. This depends
entirely upon the proportions of the vehicles employed,
and different pigments require different quantities

of vehicle. Proceed with the mixing by filling the pot two-thirds full of the dry pigment, then add boiled linseed oil, producing the consistency of putty, then dilute with about one part japan and two parts turpentine. Should it be required to keep the wood as light as possible, replace the boiled oil with raw, using a smaller quantity, but a little more japan. For all light wood the light japan should be preferred, although there are many who never use anything but the common brown. Naphtha can also be employed in place of the turpentine, somewhat reducing the expense; but, as naphtha evaporates much faster, it prevents the operator from covering an extended surface without running the risk of having it dry hard. This causes difficulty at "rubbing it in" and wiping off the surplus, although "wiping off" should not proceed until the filler has flatted—or, at least, "set." This to a certain extent the operator can accelerate or retard by omitting or adding a small quantity of oil, keeping in view the fact that the smaller the quantity of oil used, the lighter colored, but the less durable, will be the finish.

Oil is sometimes used as a filler, but its use is not recommended; applied directly to the wood, its effect is to swell the fibers, or "raise the grain," which remains in that condition until the oil becomes entirely dry or disappears. During this time the fibers are gradually shrinking, and consequently moving or checking the varnish. The qualities essential to a good filler are that it shall readily enter the porous portion of the wood, and shall very soon harden and render the wood impervious to the varnish, which should lie smoothly upon the surface, giving brilliancy and effect to the natural beauty of the wood; and that it shall not raise the grain of the wood; and that it shall not change the color of the wood. These conditions are satisfactorily fulfilled by few of the home-made fillers ordinarily used in shops, and while I give a number of recipes, my readers are advised that they will obtain better satisfaction, at less cost, by purchasing some of the patent fillers now coming into general use. In these fillers very little oil is used and a large amount of dryers, so that the wood becomes perfectly dry and hard in a few hours, preventing any swelling or shrinking of the fibers of the wood after the varnish is applied. The following fillers should be allowed to dry until quite hard. A period of about eight hours is usually sufficient, but it is better to let the work stand for twenty-four hours before touching it with sand-paper. In applying a filler it should always be borne in mind that the substance of wood consists of a multitude of small tubes lying side by side. These tubes or cells are not continuous from top to bottom of the tree, but are comparatively short and taper out to points so that they are thickest in the middle. Most of the common woods have the walls of these tubes so thin that liquid is readily absorbed by them and carried into the substance for some distance. Different kinds of wood differ much in the shape and arrangement of these cells. In filling the pores the first step is taken in providing an absolutely smooth surface. We trust mainly to mechanical force in rubbing in, aided by the absorptive powers of the wood. Formerly successive varnishings and rubbings and scrapings took much time, and when they were done, the final finish had still to be applied, but the whole process has now been simplified, by using fillers.

The careful workman will not leave "great daubs"

of superfluous filler here and there on the work, but will see that all corners and heads and quirks of mouldings are well cleaned off before it gets too hard to remove easily, and should there be any nail holes—which there ought not to be—he will have them filled with properly colored putty or cement and nicely smoothed down before he makes any attempt to put on his finishing coats.

Among the many home-made fillers I have endeavored to select the best.

Walnut Filler.—For Medium and Cheap Work. Ten lb. bolted English whiting, 3 lb. dry burnt umber, 4 lb. Vandyke brown, 3 lb. calcined plaster, ½ lb. Venetian red, 1 gal. boiled linseed oil, ½ gal. spirits turpentine, 1 quart black japan. Mix well and apply with brush; rub well with excelsior or tow, clean off with rags.

Walnut Filler.—For Imitation Wax-Finish. Five lb. bolted whiting, 1 lb. calcined plaster, 6 oz. calcined magnesia, 1 oz. dry burnt umber, 1 oz. French yellow, 1 quart raw linseed oil, 1 quart benzine spirits, ½ pint very thin white shellac. Mix well and apply with a brush. Rub well in and clean off with rags. Before using the above filling, give the work one coat of white shellac. When dry, sandpaper down and apply the filler.

Walnut Filler.—For First-Class Work. Three lb. burnt umber ground in oil, 1 lb. burnt sienna ground in oil, 1 quart spirits of turpentine, 1 pint brown japan. Mix well and apply with a brush; sandpaper well; clean off with tow and rags. This gives a beautiful chocolate color to the wood.

Filler for Light Woods.—Five lb. bolted English whiting, 3 lb. calcined plaster, 1 lb. corn-starch, 3 oz.

calcined magnesia, ½ gallon raw linseed oil, 1 quart spirits of turpentine, 1 quart brown japan, and sufficient French yellow to tinge the white. Mix well and apply with a brush, rub in with excelsior or tow, and clean off with rags.

Filler for Cherry.—Five lb. bolted English whiting, 2 lb. calcined plaster, 1½ oz. dry burnt sienna, 1 oz. Venetian red, 1 quart boiled linseed oil, 1 pint spirits of turpentine, 1 pint brown japan. Mix well, rub in with excelsior or tow and clean off with rags.

Filler for Oak.—Five lb. bolted English whiting, 2 lb. calcined plaster, 1 oz. dry burnt sienna, ½ oz. dry French yellow, 1 quart raw linseed oil, 1 pint benzine spirits, ½ pint white shellac. Mix well, apply with brush, rub in with excelsior or tow, and clean off with rag.

Filler for Rosewood.—Six lb. bolted English whiting, 2 lb. calcined plaster, 1 lb. rose pink, 2 oz. Venetian red, ½ lb. Vandyke brown, ½ lb. brandon red, 1 gallon boiled linseed oil, ½ gallon spirits of turpentine, 1 quart black japan. Mix well, apply with brush, rub in with excelsior or tow, and clean off with rags.

Another.—Stir boiled oil and corn-starch into a very thick paste; add a little japan, and reduce with turpentine, but add no color for light ash. For dark ash and chestnut use a little raw sienna; for walnut, burnt umber, add a small quantity of Venetian red; for bay wood, burnt sienna. In no case use more color than is required to overcome the white appearance of the starch, unless it is wished to stain the wood. The filler is worked with brush and rags in the usual manner. Let it dry forty-eight hours, or until it is in condition to rub down with No. 0 sandpaper without much gumming up, and if an extra fine finish is desired, fill again with the same materials, using less oil, but more of japan and turpentine.

Another.—Take three papers corn-starch, one quart boiled linseed oil, two quarts turpentine, one-quarter pint japan; cut in half the turpentine before mixing; it will not cut perfectly otherwise. For dark woods add burnt umber to color. When nearly dry, rub off with cloths. The above mixture must be used fresh, as it is of no value after it is four or five days old. The cloths used in rubbing as above mentioned should be destroyed immediately after use, as spontaneous combustion is likely to ensue from the ingredients employed. As a filler of wood to be stained, apply French plaster of Paris, mixed as a creamy paste with water, and after rubbing in, clean any surplus off; or use whiting finely powdered, or white lead slacked with painters' drying oil, and used as a filler. Another process is that of oiling, then rubbing crosswise to the grain with a sponge dipped in thin polish composed of melted beeswax, resin, and shellac, and smoothing the surface, when dry, with pumice-stone, or fine glass paper. Embody the work a second time with thicker polish, or a mixture of polish and varnish. The rubbers will work easily with half the quantity of oil which is ordinarily used. This second body should be rubbed very smooth with moist putty.

In the use of any filler, care must be taken in the selection of color, for the employment of a light colored filler on dark wood or *vice versa* would result in gross defacement, as the lighter color would show at the pores of the wood in the one case, and the darker in the other. Therefore, to avoid this, the filler should be as near as possible the color of the wood to be filled.

As a general thing, paint manufacturers who do not make fillers a specialty use opaque colors to stain their filler, as it requires a less quantity. This will do sometimes, but not always. But those which give to the wood a clear and bright appearance, and therefore produce the best results, are stained with transparent colors; those chiefly employed are burnt umber and sienna, Venetian red, Vandyke brown and charcoal black, the charcoal being ground fine in oil, while the others can be used dry and according to the following recipes with good results:

In mixing any or all dark fillers the same pigments used for the light (previously described) should be kept for a basis, with sufficient coloring to stain it to the desired depth of shade.

Filler for walnut is very often stained with burnt umber; this is reddish in hue and gives to the wood a pleasing effect. Others use Venetian red, darkened somewhat with lampblack; this is rather opaque, and tends to deaden the color of the wood. There is another article—namely, Vandyke brown—which gives fair results. In order to obtain a rich effect, the filler should be made considerably darker than the wood when new.

Fillers for mahogany, cherry, California redwood, and other woods of similar shade, should be stained with bunrt sienna, as they should be finished very clear. It is well to know that charcoal black and Venetian red will give the desired shade for any dark-colored wood in common use or for all colors in antique, but it does not show up quite as clear as some other combinations.

For rosewood, charcoal as a stain will suffice, and for vanilla or Brazil-wood the use of rose-pink will give good results.

The methods of mixing these fillers are quite numerous. It is impossible to give the proportions definitely, owing to the strength of the colors or the transparency of the chief ingredient, but one cannot go astray by following the preceding rules.

Mix the light pigment to a paste with boiled oil, which must be well stirred up. Then in another pot mix a quantity of the colored pigment with turpentine or naphtha; and when thoroughly "cut," or dissolved, add sufficient of it to the light to give the shade required. After this is obtained, dilute with turpentine or naphtha and japan, as directed in mixing light filler. This applies to all colors except black, which is seldom obtained finely ground unless in oil, and properly thinned down.

There are many finishers and firms who exclusively use manufactured fillers, and in consequence meet with many difficulties as to the shade they require, as different manufacturers use different colors to stain their filler. But this difficulty can be overcome by a few experiments with the above-named stains.

As the foregoing gives pretty nearly all the fillers in general use, with the exception of some of the manufactured mineral preparations of which I will have more to say further on, I will now proceed to describe the method of application. The secret of this is to do the work well, quickly and economically. These points are dealt with in the following:

Have your filler mixed to the consistency of ordinary lead paint; then apply to the prepared surface of the wood with a pound brush, or, what is still better, a 3-0 or 4-0 oval chisel varnish brush. In applying the filler it is not necessary to cover all the small beads and carvings; and if the filler be light, better avoid coating them at all; and if dark or antique, stain them with a little of the filler, much reduced with spirits of turpentine. For this purpose have at hand a small pot with a small fitch or sash tool.

By not filling the beads and carvings, the varnish is not so liable to run down in them, although sufficient remains to produce a finish equal to the balance of the surface.

After enough surface has been covered with the filler, so that what has been first applied begins to flatten, the process of wiping should immediately begin, using for that purpose either a rag or a handful of waste or excelsior. If the wood is very open grained, waste is preferable. With a piece of this that has previously been used and is pretty well supplied with filler, rub crosswise of the grain, rather rubbing it into the grain than wiping it off. After the whole surface has been gone over in this way, take a clean piece of waste or rag (never use excelsior for wiping clean) and wipe the surface perfectly clean and free from filler, using a wooden pick (Fig. 105), the point of which has been covered with a rag or waste, to clean out the corners, beads, etc. It is well to give these picks some attention, as a person once accustomed to certain tools can accomplish more and better work than with tools that feel strange in his hands; therefore, each finisher should furnish his own pick. As

Fig. 105

to their construction, these are best made from second-growth hickory, which can be procured at any carriage repair shop, such as old spokes, broken felloes, etc. They are made eight incnes in length, half inch oval at one end and tapering down to the point at the other. Sharpen the oval end like a cold chisel, then smooth with sandpaper, which should also be used to sharpen the tool when the same becomes worn dull.

This picking out of the filler from beads, etc., can be accelerated by the use of picking brushes, several of which I show in Fig. 106, and which are manu. factured especially for that purpose, but it is not

FIG. 106

advisable to use them on very coarsely grained wood, as they scrub the filler out of the pores.

There are several fillers used which do not require this picking and scrubbing. One is a liquid filler used chiefly for carriage finishing; but it can be used successfully on butternut, bird's-eye maple, curly maple, satinwood, hickory, etc. It is made from gum and oil. Another is a filler made from finely ground pumice-stone, mixed as other fillers. It is applied with a brush, and must be left to dry at least twenty-four hours; it is then sandpapered smooth, when an oil varnish is applied, rendering it completely transparent. This last can be used only upon light wood.

The workman, as a matter of course, will understand that different woods require slightly different treatment, and the finer-grained woods, among which are the pines, maples, cedars and poplars, of different varieties, and birch, cherry, beech, sycamore, white box, satinwood, etc., require no filling, not that a filling would prove detrimental to the finish—except upon stained work or white holly, which in order to maintain a clear color should never be filled—but, from the condition of all fine wood, it is superfluous, and only causes unnecessary labor and expense. At this point it will be convenient to pause to consider the subject of mineral or prepared wood-fillers.

A great deal of time and money have been wasted in attempting to make good fillers, to no purpose, and a great variety—as I have shown—of substances, as chalk, plaster of Paris, corn-starch, etc., etc., have been mixed with various vehicles and rubbed into the wood with but indifferent success. Most of these compounds labor under the disadvantages of forming chemical compounds with the oil and consequently they shrink very much on drying, so that though the surface may appear smooth when they are first put on, waves and hollows make their appearance as they dry. These waves, having round edges, are difficult to fill, the second coat building up as much or more upon the level spaces as in the hollows. It sometimes seems almost impossible with these fillers in the latter coats to make the hollows hold any substance, the filler clinging chiefly to the surfaces.

I have thought it necessary to show how the ordinary or home-made fillers act, and fail, in order to show by contrast how much easier it is to work efficiently with the mineral fillers. The mineral quartz, when mixed with oil, probably shrinks less in drying than any other similar known mixture. If a surface of wood be covered with this and then rubbed, the sharp and angular particles of the silica imbed themselves in the pores of the wood, closing them up, while the oil cements them fast. This is the foundation of Wheeler's wood filler, which we recommend for use by all wood finishers. When the pores have been filled with silica, and are cemented fast by the proper mixture of gums and oils, the difficult part of the work is done.

After a good surface has been made upon the article it is ready for the filler, which is to be selected according to the color desired. In putting the filler on, it is thinned with turpentine until about like flowing varnish, and is applied with a brush. Only so much of the surface is covered as can be cleaned off before it hardens. When it has set so that the gloss has left the surface, it is at once rubbed off with excelsior or shavings, going across the grain with the strokes. If the filler dries too fast or too light, a little raw linseed oil may be used in it.

Perhaps a better material for rubbing off than excelsior is hemp, or "flax tow." At any rate, the work should be finished with some finer material than excelsior.

For a nicer job the filler is rubbed in with a rubber, made by gluing a piece of sole or belt leather on the face of a block of wood and trimming the edges flush with the block. The rubbing is done after the filler has set and before it is cleaned off. If it dries off too light, a little white japan may be added on nice work.

The light-colored filler should be used on all work where light and dark woods are used together. The filling, it must be understood, is done by the silica, which will often be found in the shape of a sediment in the bottom of the mixture. Eight hours is generally considered a sufficient time for the filler to dry.

When the work with the filler is done, the surface of the wood ought to be like so much ground glass. Such portions of the wood as show a solid grain need very little filler. On Georgia pine, after the filler is dry, a little rubbing in the direction of the grain with very fine sandpaper is an advantage. If the filler has been properly used the desired results will be obtained with little labor.

The wood is now in a condition to receive the final coatings. Whether the work is to be polished or "dead finished," do not employ shellac or "French polish." If a "dead" surface is wanted, wax finish is easily put on, and as easily rubbed to a good surface. Several manufacturers in this country prepare a wax finish, which is a convenient preparation of wax and gums, and can be applied with a brush and then rubbed down with a woolen cloth, tied up to make a hard rubber, until a fine, lusterless surface is obtained. With mahogany and similar woods this greatly improves the color of the wood. When this has dried, which will be in the course of a few hours, the work is ready for use. The wax finish, like many of the furniture creams, has the advantage that it can be put on in a few minutes at any time to brighten up work when it has become dull. A piece of work prepared in this way, after four operations, will present as fine an appearance as the best cabinet work found in the furniture stores.

The materials which have been described, it will be noticed, are both manufactured articles. The prepared filler is indispensable; the wax finish can be made by mixing together, by the aid of heat, white wax and spirits of turpentine until they are of the consistency of thick paste. Another wax finish is made of bees-wax, spirits of turpentine and linseed oil in equal parts. The addition of two drams of alkanet root to every twenty ounces of turpentine darkens and enriches the color. The root is to be put into a little bag and allowed to stand in the turpentine until it is sufficiently colored.

An altogether more durable surface can be made by a little change in the treatment. When the wood is filled, instead of applying the wax, take some hard oil finish, "Luxeberry," a preparation manufactured in Detroit, Mich., and put it on with a brush precisely like varnish. The coat should not be too heavy, especially on vertical surfaces, and the brush used ought to be a good one. This material gives a most brilliant polish. By rubbing it down with a woolen cloth and pumice-stone powder it can also be made dull. Hard-oil finish does not spot with hot or cold water, is slightly elastic and is not injured by pretty severe soaking in water. It gets hard in twelve hours or less in warm weather, and overnight in winter time. It is one of the best surfaces which can be used, and has the advantage of working very well in the hands of one who is not an expert in the art of finishing wood or handling varnish. It will make a very fair surface applied direct to the unfilled wood, in which case it is a good substitute for shellac.

Wax finish has the advantage that scratches can be easily repaired without sending to the cabinetmaker or the painter. Here a word of advice to the carpenter who does any work of this character may save him some trouble and make way for the further use of the same kind of finish. When the woodwork of a house is treated in this way, be sure to leave a little bottle of the wax polish with the housekeeper, with directions as to the method of using it. In sending out a "what-not," bookcase, or any other article of similar kind, put up a little bottle of the polish and show the owner, or, preferably the lady of the house, how to repair any little scratch and make the work look "as good as new." The fresh appearance of the work will be a good advertisement, while it will prevent complaints and dissatisfaction that often follow the use of work which, when injured, cannot be restored.

It may be said that either of the methods of finishing involves a great deal of labor. This is true; but the amount is not much greater than is needed for three coats of paint, and the cost of the paint would probably be more than the cost of the finish. The labor in one case can be of a cheap character, and in the other an experienced painter must be employed. The profit upon the "dead finish" can go into the pocket of the carpenter, while that of the painting must in any event be divided between the carpenter and painter, or belong to the latter altogether, who is, after all, the proper person to do the work.

I have now said about all that is necessary in the matter of "fillers" and "filling," but, as it sometimes happens that the old system of "sizing" has to be resorted to for certain kinds of work, I give herewith a formula for its construction and use:

Size of different kinds is sometimes applied to the

surface of wood to prevent absorption of the varnish. The kind of material used for the size is not important, the object being only to prevent absorption by a very thin coat of some substance not soluble in the varnish. For dark-colored wood, thin size, made by reducing ordinary glue with water, is generally used; but for lighter-colored surfaces a white size is used, which is prepared by boiling white kid or other leather or parchment-cuttings in water for a few hours, or until it forms a thin, jelly-like substance, which is reduced with water to a thin consistency, and used in a tepid state. Sometimes solutions of isinglass or tragacanth are employed in like manner. Unlike the best fillers, sizes of any kind do not improve the finish, and are sometimes a positive detriment to it. They are used solely as an economy to reduce the quantity of the varnish needed; and their use is not recommended for the best work.

WOOD-STAINING GENERALLY

There are many cases where an article constructed of wood may be more conveniently and suitably finished by staining and polishing than by painting. The practice of staining woods is much less common in America and England than on the Continent, where workmen, familiar with the different washes, produce the most delicate tones of color and shade. Wood is often stained to imitate darker and dearer varieties, but more legitimately to improve the natural appearance by heightening and bringing out the original markings, or by giving a definite color without covering the surface and hiding the nature of the material by coats of paint. The best woods for staining are those of close, even texture, as pear and cherry,

birch, beech, and maple, though softer and coarser
kinds may be treated with good effect. The wood
should be dried, and if an even tint is desired, its
surface planed and sandpapered. All the stains
should, if possible, be applied hot, as they thus penetrate more deeply into the pores. If the wood is to be
varnished, and not subjected to much handling, almost
any of the brilliant mordants used in wool and cotton
dyeing may be employed in an alcoholic solution;
but when thus colored it has an unnatural appearance,
and is best used on small surfaces only, for inlaying,
etc. The ebonized wood, of late years so much in
vogue, is in many respects the most unsatisfactory of
the stains, as the natural character and markings are
completely blotted out, and it shows the least scratch
or rubbing. Sometimes, in consequence of the quality
of the wood under treatment, it must be freed from its
natural colors by a preliminary bleaching process. To
this end it is saturated as completely as possible with
a clear solution of 17¼ oz. chloride of lime and 2 oz.
soda crystals, in 10½ pints water. In this liquid the
wood is steeped for ½ hour, if it does not appear to
injure its texture. After this bleaching, it is immersed
in a solution of sulphurous acid to remove all traces
of chlorine, and then washed in pure water. The
sulphurous acid, which may cling to the wood in spite
of washing, does not appear to injure it, nor alter the
colors which are applied.

Black.—(1) Obtained by boiling together blue
Brazil-wood, powdered gall-apples, and alum, in rain
or river water, until it becomes black. This liquid is
then filtered through a fine organdie, and the objects
painted with a new brush before the decoction has
cooled, and this repeated until the wood appears of

a fine black color. It is then coated with the following liquid: A mixture of iron filings, vitriol, and vinegar is heated (without boiling), and left a few days to settle. Even if the wood is black enough, yet, for the sake of durability, it must be coated with a solution of alum and nitric acid, mixed with a little verdigris; then a decoction of gall-apples and logwood dyes is used to give it a deep black. A decoction may be made of brown Brazil-wood with alum in rain-water, without gall-apples; the wood is left standing in it for some days in a moderately warm place, and to it merely iron filings in strong vinegar are added, and both are boiled with the wood over a gentle fire. For this purpose soft pear-wood is chosen, which is preferable to all others for black staining.

(2) 1 oz. nut-gall broken into small pieces, put into barely ½ pint vinegar, which must be contained in an open vessel; let stand for about ½ hour; add 1 oz. steel filings; the vinegar will then commence effervescing; cover up, but not sufficient to exclude all air. The solution must then stand for about 2½ hours, when it will be ready for use. Apply the solution with a brush or piece of rag to the article, then let it remain until dry; if not black enough, coat it until it is—each time, of course, letting it remain sufficiently long to dry thoroughly. After the solution is made, keep it in a closely corked bottle.

(3) One gal. water, 1 lb. logwood chips, ½ lb. black copperas, ½ lb. extract of logwood, ½ lb. indigo blue, 2 oz. lampblack. Put these into an iron pot and boil them over a slow fire. When the mixture is cool, strain it through a cloth, add ¼ oz. nut-gall. It is then ready for use. This is a good black for all kinds of cheap work.

(4) Two hundred fifty parts of Campeachy wood, 2000 water, and 30 copper sulphate; the wood is allowed to stand 24 hours in this liquor, dried in the air, and finally immersed in iron nitrate liquor at 4° B.

(5) Boil 8¾ oz. logwood in 70 oz. water and 1 oz. blue stone, and steep the wood for 24 hours. Take out, expose to the air for a long time, and then steep for 12 hours in a solution of iron nitrate at 4° B. If the black is not fine, steep again in the logwood liquor.

(6) It is customary to employ the clear liquid obtained by treating 2 parts powdered galls with 15 parts wine, and mixing the filtered liquid with a solution of iron protosulphate. Reimann recommends the use of water in the place of wine.

(7) Almost any wood can be dyed black by the following means: Take logwood extract such as is found in commerce, powder 1 oz., and boil it in 3¼ pints of water; when the extract is dissolved, add 1 dr. potash yellow chromate (not the bichromate), and agitate the whole. The operation is now finished, and the liquid will serve equally well to write with or to stain wood. Its color is a very fine dark purple, which becomes a pure black when applied to the wood.

(8) For black and gold furniture, procure 1 lb. logwood chips, add 2 qt. water, boil 1 hour, brush the liquor in hot, when dry give another coat. Now procure 1 oz. green copperas, dissolve it in warm water, well mix, and brush the solution over the wood; it will bring out a fine black; but the wood should be dried outdoors, as the black sets better. A common stove brush is best. If polish cannot be used, proceed as follows: Fill up the grain with black glue—i.e., thin glue and lampblack—brushed over the parts accessible (not in the carvings); when dry, paper down

with fine paper. Now procure, say, a gill of French polish, in which mix 1 oz. best ivory black, or gas-black is best, well shake it until quite a thick pasty mass, procure ½ pint brown hard varnish, pour a portion into a cup, add enough black polish to make it quite dark, then varnish the work; two thin coats are better than one thick coat. The first coat may be sandpapered down where accessible, as it will look better. A coat of glaze over the whole gives a piano finish. N.B.—Enough varnish should be mixed at once for the job to make it all one color—i.e., good black.

(9) For table. Wash the surface of table with liquid ammonia, applied with a piece of rag; the varnish will then peel off like a skin; afterwards smooth down with fine sandpaper. Mix ¼ lb. lampblack with 1 qt. hot water, adding a little glue size; rub this stain well in; let it dry before sand-papering it; smooth again. Mind you do not work through the stain. Afterwards apply the following black varnish with a broad, fine camel-hair brush: Mix a small quantity of gas-black with the varnish. If one coat of varnish is not sufficient, apply a second one after the first is dry. Gas-black can be obtained by boiling a pot over the gas, letting the pot nearly touch the burner, when a fine jet black will form on the bottom, which remove, and mix with the varnish. Copper vessels give the best black; it may be collected from barbers' warming pots.

(10) Black-board wash, or "liquid slating."—(a) Four pints 95 per cent alcohol, 8 oz. shellac, 12 dr. lampblack, 20 dr. ultramarine blue, 4 oz. powdered rotten stone, 6 oz. powdered pumice. (b) 1 gal. 95 per cent alcohol, 1 lb. shellac, 8 oz. best ivory black, 5 oz.

finest flour emery, 4 oz. ultramarine blue. Make a perfect solution of the shellac in the alcohol before adding the other articles. To apply the slating, have the surface smooth and perfectly free from grease; well shake the bottle containing the preparation, and pour out a small quantity only into a dish, and apply it with a new flat varnish brush as rapidly as possible. Keep the bottle well corked, and shake it up each time before pouring out the liquid. (c) Lampblack and flour of emery mixed with spirit varnish. No more lampblack and flour of emery should be used than are sufficient to give the required black abrading surface. The thinner the mixture the better. Lampblack should first be ground with a small quantity of spirit varnish or alcohol to free it from lumps. The composition should be applied to the smoothly planed surface of a board with a common paint brush. Let it become thoroughly dry and hard before it is used. Rub it down with pumice if too rough. (d) ½ gal. shellac varnish, 5 oz. lampblack, 3 oz. powdered iron ore or emery; if too thick, thin with alcohol. Give 3 coats of the composition, allowing each to dry before putting on the next; the first may be of shellac and lampblack alone. (e) To make 1 gal. of the paint for a blackboard, take 10 oz. pulverized and sifted pumice, 6 oz. powdered rotten stone (infusorial silica), ¾ lb. good lampblack, and alcohol enough to form with these a thick paste, which must be well rubbed and ground together. Then dissolve 14 oz. shellac in the remainder of the gallon of alcohol by digestion and agitation, and finally mix this varnish and the paste together. It is applied to the board with a brush, care being taken to keep the paint well stirred, so that the pumice will not settle. Two coats are usually necessary

The first should be allowed to dry thoroughly before the second is put on, the latter being applied so as not to disturb or rub off any portion of the first. One gallon of this paint will ordinarily furnish 2 coats for 60 sq. yd. of blackboard. When the paint is to be put on plastered walls, the wall should be previously coated with glue size—1 lb. glue, 1 gal. water, enough lampblack to color; put on hot. (f) Instead of the alcohol mentioned in b, take a solution of borax in water; dissolve the shellac in this and color with lampblack. (g) Dilute soda silicate (water-glass) with an equal bulk of water, and add sufficient lampblack to color it. The lampblack should be ground with water and a little of the silicate before being added to the rest of the liquid

(11) 17.5 oz. Brazil-wood and 0.525 oz. alum are boiled for 1 hour in 2.75 lb. water. The colored liquor is then filtered from the boiled Brazil-wood, and applied several times boiling hot to the wood to be stained. This will assume a violet color. This violet color can be easily changed into black by preparing a solution of 2.1 oz. iron filings, and 1.05 oz. common salt in 17.5 oz. vinegar. The solution is filtered, and applied to the wood, which will then acquire a beautiful black color.

(12) 8.75 oz. gall-nuts and 2.2 lb. logwood are boiled in 2.2 lb. rain-water for 1 hour in a copper boiler. The decoction is then filtered through a cloth, and applied several times while it is still warm to the article of wood to be stained. In this manner a beautiful black will be obtained.

(13) This is prepared by dissolving 0.525 oz. logwood extract in 2.2 lb. hot rain-water, and by adding to the logwood solution 0.035 oz. potash chromate.

When this is applied several times to the article to be stained, a dark brown color will first be obtained. To change this into a deep chrome-black, the solution of iron filings, common salt, and vinegar, given under (11) is applied to the wood, and the desired color will be produced.

(14) Several coats of alizarine ink are applied to the wood, but every coat must be thoroughly dry before the other is put on. When the articles are dry, the solution of iron filings, common salt, and vinegar, as given in (11), is applied to the wood, and a very durable black will be obtained.

(15) According to Herzog, a black stain for wood, giving to it a color resembling ebony, is obtained by treating the wood with two fluids, one after the other. The first fluid to be used consists of a very concentrated solution of logwood, and to 0.35 oz. of this fluid are added 0.017 oz. alum. The other fluid is obtained by digesting iron filings in vinegar. After the wood has been dipped in the first hot fluid, it is allowed to dry, and is then treated with the second fluid, several times if necessary.

(16) Sponge the wood with a solution of aniline chlorhydrate in water, to which a small quantity of copper chloride is added. Allow it to dry, and go over it with a solution of potassium bichromate. Repeat the process two or three times, and the wood will take a fine black color.

Blue.—(1) Powder a little Prussian blue, and mix to the consistency of paint with beer; brush it on the wood, and when dry size it with glue dissolved in boiling water; apply lukewarm, and let this dry also; then varnish or French polish.

(2) Indigo solution, or a concentrated hot solution

of blue vitriol, followed by a dip in a solution of washing soda.

(3) Prepare as for violet, and dye with aniline blue.

(4) A beautiful blue stain is obtained by gradually stirring 0.52 oz. finely powdered indigo into 4.2 oz. sulphuric acid of 60 per cent, and by exposing this mixture for 12 hours to a temperature of 77° F. (25° C.). The mass is then poured into 11-13.2 lb. rain-water, and filtered through felt. This filtered water is applied several times to the wood, until the desired color has been obtained. The more the solution is diluted with water, the lighter will be the color.

(5) 1.05 oz. finest indigo carmine, dissolved in 8.75 oz. water, applied several times to the articles to be stained. A very fine blue is in this manner obtained.

(6) 3.5 oz. French verdigris are dissolved in 3.5 oz. urine and 8.75 oz. wine vinegar. The solution is filtered and applied to the article to be stained. Then a solution of 2.1 oz. potash carbonate in 8.75 oz. rain-water is prepared, and the article colored with the verdigris is brushed over with this solution until the desired blue color makes its appearance.

(7) The newest processes of staining wood blue are those with aniline colors. The following colors may be chosen for the staining liquor: Bleu de Lyon (reddish blue), bleu de lumiere (pure blue), light blue (greenish blue). These colors are dissolved in the proportion of 1 part coloring substance to 30 of spirit of wine, and the wood is treated with the solution.

Brown.—(1) Various tones may be produced by mordanting with potash chromate, and applying a decoction of fustic, of logwood, or of peachwood.

(2) Sulphuric acid, more or less diluted according to the intensity of the color to be produced, is applied

with a brush to the wood, previously cleaned and dried. A lighter or darker brown stain is obtained, according to the strength of the acid. When the acid has acted sufficiently, its further action is arrested by the application of ammonia.

(3) Tincture of iodine yields a fine brown coloration, which, however, is not permanent unless the air is excluded by a thick coating of polish.

(4) A simple brown wash is ½ oz. alkanet root, 1 oz. aloes, 1 oz. dragons' blood, digested in 1 lb. alcohol. This is applied after the wood has been washed with aqua regia, but is, like all the alcoholic washes, not very durable.

Ebonizing.—(1) Boil 1 lb. logwood chips 1 hour in 2 qt. water; brush the hot liquor over the work to be stained, lay aside to dry; when dry give another coat, still using it hot. When the second coat is dry, brush the following liquor over the work: 1 oz. green copperas to 1 qt. hot water, to be used when the copperas is all dissolved. It will bring out an intense black when dry. For staining, the work must not be dried by fire, but in the sunshine, if possible; if not, in a warm room, away from the fire. To polish this work, first give a coating of very thin glue size, and when quite dry paper off very lightly with No. 0 paper, only just enough to render smooth, but not to remove the black stain. Then make a rubber of wadding about the size of a walnut, moisten the rubber with French polish, cover the whole tightly with a double linen rag, put one drop of oil on the surface, and rub the work with a circular motion. Should the rubber stick it requires more polish. Previous to putting the French polish on the wadding pledget, it ought to be mixed with the best drop black, in the proportion of

¼ oz. drop black to a gill of French polish. When the work has received one coat, set it aside to dry for about an hour. After the first coat is laid on and thoroughly dry, it should be partly papered off with No. 0 paper. This brings the surface even, and at the same time fills up the grain. Now give a second coat as before. Allow 24 hours to elapse, again paper off, and give a final coat as before. Now comes "spiriting off." Great care must be used here, or the work will be dull instead of bright. A clean rubber must be made, as previously described, but instead of being moistened with polish it must be wetted with spirits of wine placed in a linen rag screwed into a tight, even-surfaced ball, just touched on the face with a drop of oil, and then rubbed lightly and quickly in circular sweeps all over the work from top to bottom. One application of spirits is usually enough if sufficient has been placed on the rubber at the outset, but it is better to use rather too little than too much at a time, as an excess will entirely remove the polish, when the work will have to be polished again. Should this be the case, paper off at once, and commence as at first. It is the best way in the end. (*Smither.*)

(2) Lauber dissolves extract of logwood in boiling water until the solution indicates 0° Beaumé. Five pints of the solution is then mixed with 2½ pints pyroligneous iron mordant of 10°, and ½ pint acetic acid of 2°. The mixture is heated for ¼ hour, and is then ready for use.

(3) To imitate black ebony, first wet the wood with a solution of logwood and copperas, boiled together and laid on hot. For this purpose, 2 oz. logwood chips with 1½ oz. copperas, to 1 qt. water, will be required. When the work has become dry, wet the

surface again with a mixture of vinegar and steel filings. This mixture may be made by dissolving 2 oz. steel filings in ½ pint vinegar. When the work has become dry again, sandpaper down until quite smooth. Then oil and fill in with powdered drop black mixed in the filler. Work to be ebonized should be smooth and free from holes, etc. The work may receive a light coat of quick-drying varnish, and then be rubbed with finely pulverized pumice and linseed oil until very smooth.

(4) One gal. strong vinegar, 2 lb. extract of logwood, ½ lb. green copperas, ¼ lb. China blue, and 2 oz. nut-gall. Put these in an iron pot, and boil them over a slow fire till they are well dissolved. When cool, the mixture is ready for use. Add to the above ½ pint iron rust, which may be obtained by scraping rusty hoops, or preferably by steeping iron filings in a solution of acetic acid or strong vinegar.

(5) Common ebony stain is obtained by preparing two baths; the first, applied warm, consists of a logwood decoction, to every quart of which 1 dr. alum is added; the second is a solution of iron filings in vinegar. After the wood has dried from the first, the second is applied as often as is required. For the first-named bath, some substitute 16 oz. gall-nut, 4 oz. logwood dust, and 2 oz. verdigris, boiled in a sufficient quantity of water. A peculiar method of blackening walnut is in use in Nürnberg. On one of the Pegnitz Islands there is a large grinding-mill, turned by the stream, where iron tools are sharpened and polished. The wood is buried for a week or more in the slime formed by the wheels; when dug out it is jet black, and so permeated by silica as to be in effect petrified. Another way to ebonize flat surfaces of soft

work is to rub very fine charcoal dust into the pores
with oil. This works beautifully with basswood and
American whitewood. A brown mahogany-like stain
is best used on elm and walnut. Take a pint decoction
of 2 oz. logwood in which ½ oz. barium chloride has
been dissolved. This gives also, when diluted with
soft water, a good oak stain to ash and chestnut. But
the most beautiful and lasting of the browns is a concentrated solution of potash permanganate (mineral
chameleon). This is decomposed by the woody fiber,
and forms hydrated manganese oxide, which is
permanently fixed by the alkali.

(6) For the fine black ebony stain, apple, pear, and
hazel wood are the best woods to use; when stained
black, they are most complete imitations of the
natural ebony. For the stain take: gall-apple, 14 oz.;
rasped logwood, 3½ oz.; vitriol, 1¾ oz.; verdigris,
1¾ oz. For the second coating a mixture of iron
filings (pure), 3½ oz., dissolved in strong wine vinegar;
1½ pint is warmed, and when cool the wood already
blackened is coated 2 or 3 times with it, allowing it
to dry after each coat. For articles which are to
be thoroughly saturated, a mixture of 1¾ oz. sal-
ammoniac, with a sufficient quantity of steel filings, is
to be placed in a suitable vessel, strong vinegar poured
upon it, and left for 14 days in a gently heated oven.
A strong lye is now put into a suitable pot, to which
is added coarsely bruised gall-apples and blue Brazil
shavings, and exposed for the same time as the former
to the gentle heat of an oven, which will then yield a
good liquid. The woods are now laid in the first-
named stain, boiled for a few hours, and left in it for 3
days longer; they are then placed in the second stain
and treated as in the first. If the articles are not then

thoroughly saturated, they may be once more placed
in the first bath, and then in the second. The polish
used for wood that is stained black should be "white"
(colorless) polish, to which a very little finely ground
Prussian blue should be added.

(7) Wash with a concentrated aqueous solution of
logwood extract several times; then with a solution
of iron acetate of 14° B., which is repeated until a deep
black is produced.

(8) Beech, pear-tree, or nolly steeped in a strong
liquor of logwood or galls. Let the wood dry, and
wash over with solution of iron sulphate. Wash with
clean water, and repeat if color is not dark enough.
Polish either with black or common French polish.

(9) Oak is immersed for 48 hours in a hot saturated
solution of alum, and then brushed over several times
with a logwood decoction prepared as follows: Boil 1
part best logwood with 10 of water, filter through linen,
and evaporate at a gentle heat until the volume is
reduced one-half. To every quart of this add 10 to 15
drops of a saturated solution of indigo, completely
neutral. After applying this dye to the wood, rub
the latter with a saturated and filtered solution of
verdigris in hot concentrated acetic acid, and repeat
the operation until a black of the desired intensity is
obtained. Oak thus stained is said to be a close as
well as handsome imitation of ebony.

(10) One lb. logwood chips, 3 pints water; boil to 1
pint; apply hot to wood; let dry; then give another
coat; let dry slowly; sandpaper smooth; mix 1 gill
vinegar with 3 tablespoonfuls iron or steel filings; let
stand 5 hours, then brush on wood; let dry; then give
another coat of the first. This sends the vinegar
deeper into the wood and makes a denser black; after

which paper smooth. Then polish with white French polish, as the white brings out the black purer than common French polish. The woods observed to take on the stain best are pear-tree, plane-tree, and straight-reeded birch; mahogany does not stain nearly so well as the former woods.

(11) Get 1 lb. logwood chips and boil them down in enough water to make a good dark color; give the furniture 3 or 4 coats with a sponge; then put some rusty nails or old iron into a bottle with some vinegar, and when it begins to work give the furniture a coat of the vinegar. This, if you have well darkened it with the first, will give you a good black. Oil and polish in the usual way, rubbing down first with fine paper if required. A quicker way is to give the wood a coat of size and lampblack, and then use gas-black in your polish rubber

(12) Make a strong decoction of logwood by boiling 1 lb. in 1 qt. water for about 1 hour; add thereto a piece of washing soda as large as a hazel-nut. Apply hot to the wood with a soft brush. Allow to dry, then paint over the wood with a solution of iron sulphate (1 oz. to the pint of water). Allow this to dry, and repeat the logwood and iron sulphate for at least 3 times, finishing off with logwood. Once more allow to dry thoroughly, then sandpaper off very lightly (so as not to remove the dye) with No. 0 paper. Now make a very thin glue size, boil in it a few chips of logwood and a crystal or two of iron sulphate, just sufficient to make it inky black. Paint this lightly over the work, allow to dry once more, again sandpaper lightly, and finally either varnish with good hard white varnish, or polish with French polish and drop black.

Floors. — (1) Get the wood clean, have some Vandyke brown and burnt sienna ground in water, mix it in strong size, put on with a whitewash or new paint brush as evenly as you can. When dry, give 2 coats of copal or oak varnish.

(2) If the floor is a new one, have the border well washed. Polish with sandpaper, rubbing always with the grain of the wood. Varnish with good oak varnish, put coloring matter into the varnish to suit your taste, but umber is best; if the floor is old and blackened, paint it.

(3) If old floors, you will not make much of staining anything but black. The floor is to be well washed (lime and soda is best—no soap), the dye painted on, and, when dry, sized over and varnished with elastic oak varnish.

(4) Take ½ lb. logwood chips, boil them briskly for ½ hour in about 5 qt. rain-water, and strain through muslin. To this liquor add 6 oz. annatto (in the form of cake—not the roll); add also 1 lb. of yellow wax cut up in very small pieces. Place these over the fire, and let the wax melt, gently stirring it all the while. When melted, take the mixture off the fire; do not let it boil. Then with a paint brush lay it on the floor as hot as possible, brushing it always the way of the grain. Next day polish with a hard, flat brush made of hair, which may have a strap nailed to the back of it in which to insert the foot. The floor is afterwards kept bright with beeswax alone, a little of which is melted and put on the brush. Take care that the floor is thoroughly dry before commencing operations.

(5) Melt some glue size in a bottle; next get a piece of rag, roll it into a ball so that it will fit the hand

nicely, cover this with a bit of old calico to make a smooth face; dip this into the size, and rub in a bit of brown umber; then go ahead with your floors, working the stuff light or dark as required. Keep the motion with the grain of wood; when dry, stiffen with polishers' glaze.

(6) Take Judson's dyes of the color required, mix according to the instructions given with each bottle, and apply with a piece of rag, previously trying it on a piece of wood to see if color would suit; rub with sandpaper to get off any roughness that may be raised with the damp, and varnish with fine, pale hard varnish, then slightly sandpaper and varnish again. Another method is to boil 1 lb. logwood in an *old* boiler, then apply with a piece of rag where the stain is required; when thoroughly dry, sandpaper as before, and well rub with beeswax to polish. This last process looks best when finished, but it requires a lot of elbow grease for a few months, and is extremely durable. To prevent the stain running where you do not want it, paste some stout paper.

(7) As a general rule, 1 qt. of the staining liquid will be found sufficient to cover about 16 sq. yd. of flooring; but different kinds of woods absorb in different proportions, soft woods requiring more for the same space than hard woods. The colors of the stains are various, so that one may either choose ebony, walnut, mahogany, rosewood, satinwood, oak, medium oak, or maple, according to the paleness or depth of color desired. Besides this, 4 lb. of size and 2½ pints of the best varnish are required to finish the 16 yd. above mentioned. The necessary purchases are completed by a good-sized painters' brush and a smaller one. The work can then be commenced. It

the wood is uneven, it must be planed, and rubbed
down to a smooth surface; whilst the cracks and spaces
between the boards, if very wide, may be disposed of
by a process called "slipping," by which pieces of
wood are fitted in. The floor must next be carefully
washed, and allowed to dry thoroughly. The actual
staining may now be proceeded with. The liquid is
poured out into a basin, and spread all over the floor
with the aid of the large brush, the small one being
used to do the corners and along the wainscoting, so
that it may not be smeared. It is always best to begin
staining at the farthest corner from the doorway, and
work round so that one's exit may not be impeded.
It is also a good plan to work with the window open,
if there is no danger of much dust flying in, as the
staining dries so much quicker. After the floor is quite
covered, the stainer may rest for about an hour whilst
the drying is going on, during which there is only one
thing relative to the work in hand which need be
attended to. This is the size, which should be put in
a large basin with ½ pint of cold water to each pound,
and then stood in a warm place to dissolve. Before
recommencing work, also, the brushes must be washed,
and this is no great trouble, as a little lukewarm water
will take out all trace of the stain and clean them
quite sufficiently. The sizing is then laid on in
exactly the same manner as the staining, always being
careful to pass the brush lengthwise down the boards.
If the size froths or sticks unpleasantly, it must be a
little more diluted with warm water, and sometimes, if
the sediment from it is very thick, it is all the better
for being strained through a coarse muslin. The
sizing takes rather longer than the varnish to dry, 2 or
more hours being necessary, even on a warm, dry day.

Not until it is quite dry, however, can the last finish be put to the work with the varnish. For this it is always safest to get the very best, and to lay it on rather literally, though very evenly, and over every single inch, as the staining will soon rub off when not protected by it. The best way to ascertain whether it is varnished all over is to kneel down and look at the floor sideways, with one's eyes almost on a level with it.

Green.—(1) Mordant the wood with red liquor at 1° B. This is prepared by dissolving separately in water 1 part sugar of lead and 4 of alum free from iron; mix the solutions, and then add $\frac{1}{32}$ part of soda crystals, and let settle overnight. The clear liquor is decanted off from the sediment of lead sulphate, and is then diluted with water till it marks 1° B. The wood when mordanted is dyed green with berry liquor and indigo extract, the relative proportions of which determine the tone of the green.

(2) Verdigris dissolved in 4 parts water.

(3) 4.2 oz. copper, cut up finely, are gradually dissolved in 13 oz. nitric acid (aqua fortis), and the articles to be stained are boiled in this solution until they have assumed a fine green color.

Gray.—(1) Grays may be produced by boiling 17 oz. orchil paste for ½ hour in 7 pints water. The wood is first treated with this solution, and then, before it is dry, steeped in a beck of iron nitrate at 1° B. An excess of iron gives a yellowish tone; otherwise a blue gray is produced, which may be completely converted into blue by means of a little potash.

(2) One part silver nitrate dissolved in 50 of distilled water; wash over twice; then with hydrochloric acid and afterwards with water of ammonia.

The wood is allowed to dry in the dark, and then finished in oil and polished.

Mahogany.—(1) Boil ½ lb. madder and 2 oz. logwood chips in 1 gal. water, and brush well over while hot. When dry, go over with pearlash solution, 2 dr. to the quart. By using it strong or weak, the color can be varied at pleasure.

(2) Soak 1 lb. stick varnish in 2 qt. water until all the color is dissolved out; strain off the water, and add to the residue 25 dr. powdered madder. Set the mixture over the fire until it is reduced to ¾ of its original volume. Then mix together 25 dr. cochineal, 25 dr. kermes berries, 1 pint spirits of wine, and ½ oz. pearlash, out of which the color has been washed by soaking in a gill of soft water. Add this mixture to the decoction of madder and varnish, stirring well together, and adding so much aqua fortis as will bring the red to the desired shade.

(3) DARK MAHOGANY.—Introduce into a bottle 15 gr. alkanet root, 30 gr. aloes, 30 gr. powdered dragons' blood, and 500 gr. 95 per cent alcohol, closing the mouth of the bottle with a piece of bladder, keeping it in a warm place for 3 or 4 days, with occasional shaking, then filtering the liquid. The wood is first mordanted with nitric acid, and when dry washed with the stain once or oftener, according to the desired shade; then, the wood being dried, it is oiled and polished.

(4) LIGHT MAHOGANY.—Same as dark mahogany, but the stain being only applied once. The veins of true mahogany may be imitated by the use of iron acetate skillfully applied.

(5) The following process is recommended in "Wiederhold's Trade Circular:"—The coarse wood is

first coated with a colored size, which is prepared by thoroughly mixing up, in a warm solution, 1 part commercial glue in 6 of water, a sufficient quantity of the commercial mahogany brown, which is in reality an iron oxide, and in color stands between so-called Indian red and iron oxide. This is best effected by adding in excess a sufficient quantity of the dry color with the warm solution of glue, and thoroughly mixing the mass by means of a brush until a uniform paste is obtained, in which no more dry red particles are seen. A trial coat is then laid upon a piece of wood. If it is desired to give a light mahogany color to the object, it is only necessary to add less, and, for a darker color, more, of the brown body-color. When the coat is dry, it may be tested, by rubbing with the fingers, whether the color easily separates or not. In the former case, more glue must be added until the dry trial coat no longer perceptibly rubs off with the hands. Having ascertained in this way the right condition of the size color with respect to tint and strength, it is then warmed slightly, and worked through a hair sieve by means of a brush. After this, it is rubbed upon the wood surface with the brush, which has been carefully washed. It is not necessary to keep the color warm during the painting. Should it become thick by gelatinizing, it may be laid on the wood with the brush, and dries more rapidly than when the color is too thin. If the wood is porous and absorbs much color, a second coat may be laid on the first when dry, which will be sufficient in all cases. On drying, the size color appears dull and unsightly, but the following coat changes immediately the appearance of the surface. This coat is spirit varnish. For its production 3 parts spirits of wine of 90° are added in excess to 1 part of red acaroid resin in one vessel, and in another 10 parts shellac with 40 of spirits of wine of 80%. By repeated agitation for 3 or 4 days, the spirit dissolves the resin completely. The shellac solution is then poured carefully from the sediment, or, better still, filtered through a fine cloth, when it may be observed that a slight milky turbidity is no detriment to its use. The resin solution is best filtered into the shellac solution by pouring through a funnel loosely packed with wadding. When filtered, the solutions of both resins are mixed by agitating the vessel and letting the varnish stand a few days. The acaroid resin colors the shellac, and imparts to it at the same time the degree of suppleness usually obtained by the addition of Venetian turpentine or linseed oil. If the varnish is to be employed as a coat, the upper layers are poured off at once from the vessel. One or two coats suffice, as a rule, to give the object an exceedingly pleasing effect. The coats dry very quickly, and care must be taken not to apply the second coat until the first is completely dry.

(6) 7.5 oz. madder, 8.75 oz. rasped yellow wood, are boiled for 1 hour in 5.5 lb. water, and the boiling liquor is applied to the articles until the desired color has been produced.

(7) 1.05 oz. powdered turmeric, 1.05 oz. powdered dragons' blood, are digested in 8.75 oz. of 80 per cent strong alcohol, and when the latter seems to be thoroughly colored it is filtered through a cloth. The filtrate is heated and applied warm to the article.

(8) 17.5 oz. madder, 8.75 oz. ground logwood, are boiled for 1 hour in 5.5 lb. water. This is filtered while still warm, and the warm liquor is applied to the wood. When this has become dry, and it is

desired to produce a darker mahogany color, a solution of 0.525 oz. potash carbonate in 4.4 lb. water is applied to the wood. This solution is prepared cold, and filtered through blotting-paper.

(9) 0.35 oz. aniline is dissolved in 8.75 oz. spirits of wine 90 per cent strong. Then another solution of 0.35 oz. aniline yellow in 17.5 oz. spirits of wine 90 per cent strong is made, and this is added to the aniline solution until the required reddish-yellow color is obtained. By adding a little of a solution of aniline brown (0.35 oz. aniline brown in 10.5 oz. spirits of wine 90 per cent strong), the color is still more completely harmonized, and a tint very closely resembling mahogany can be given to elm and cherry wood with this mixture.

(10) 0.7 oz. logwood is boiled in 3.5 oz. water down to about ½. This is then filtered, and 0.12 oz. baryta chloride is dissolved in it.

Oak.—(1) Mix powdered ocher, Venetian red, and umber, in size, in proportions to suit; or a richer stain may be made with raw sienna, burnt sienna, and Vandyke. A light yellow stain of raw sienna alone is very effective.

(2) Darkening Oak.—Lay on liquid ammonia with a rag or brush. The color deepens immediately, and does not fade; this being an artificial production of the process which is induced naturally by age. Potash bichromate, dissolved in cold water and applied in a like manner, will produce a very similar result.

(3) In Germany, the cabinetmakers use very strong coffee for darkening oak. To make it very dark: iron filings with a little sulphuric acid and water, put on with a sponge, and allowed to dry between each application until the right hue is reached.

(4) Whitewash with fresh lime, and when dry brush off the lime with a hard brush, and dress well with linseed oil. It should be done after the wood has been worked, and it will make not only the wood, but the carving or moulding, look old also.

(5) Use a strong solution of common washing-soda, say one or two coats, until the proper color is obtained. Or you may try potash carbonate. Paper and finish off with linseed oil.

(6) A decoction of green walnut-shells will bring new oak to any shade, or nearly black.

(7) A good method of producing the peculiar olive brown of old oak is by fumigation with liquid ammonia; the method has many advantages beyond the expense of making a case or room airtight and the price of the ammonia. It does not raise the grain, the work keeping as smooth as at first. Any tint, or rather, depth of the color can be given with certainty; and the darker shade of color will be found to have penetrated to the depth of a veneer, and much farther where the end grain is exposed, thus doing away with the chance of an accidental knock showing the white wood. The coloring is very even and pure, not destroying the transparency of the wood. It is advisable to make the furniture from one kind of stuff, not to mix white oak with red, and so on. They both take the color well, but there is a kind of American live oak that does not answer well. In all cases care must be taken to have no glue or grease on the work, which would cause white spots to be left. The deal portions of the work are not affected in the least, neither does it affect the sap of oak. The best kind of polish for furniture treated in this manner is wax polish, or the kind known as egg-shell polish. The

process of fumigation is very simple. Get a large packing case, or better still, make a room in a corner of the polishing shop about 9 ft. long, 6 ft. high, and 3 ft. 6 in. wide; pass paper over the joints; let the door close on to a strip of india-rubber tubing; put a pane of glass in the side of box or house to enable you to examine the progress of coloring. In putting in your work see that it does not touch anything to hinder the free course of the fumes. Put 2 or 3 dishes on the floor to hold the ammonia; about ½ pint is sufficient for a case this size. The ammonia differs in purity, some leaving more residue than others. Small articles can be done by simply covering them with a cloth, having a little spirits in a pot underneath. A good useful color can be given by leaving the things exposed to the fumes overnight. The color lightens on being polished, owing to the transparency thus given to the wood.

Purple.—(1) Take 1 lb. logwood chips, ¾ gal. water, 4 oz. pearlash, 2 oz. powdered indigo. Boil the logwood in the water till the full strength is obtained, then add the pearlash and indigo, and when the ingredients are dissolved the mixture is ready for use, either warm or cold. This gives a beautiful purple.

(2) To stain wood a rich purple or chocolate color, boil ½ lb. madder and ¼ lb. fustic in 1 gal. water, and when boiling brush over the work until stained. If the surface of the work should be perfectly smooth, brush over the work with a weak solution of nitric acid; then finish with the following: put 4½ oz. dragons' blood and 1 oz. soda, both well bruised, into 3 pints spirits of wine. Let it stand in a warm place, shake frequently, strain and lay on with a soft brush,

repeating until a proper color is gained. Polish with linseed oil or varnish.

(3) 2.2 lb. rasped logwood, 5.5 lb. rasped Lima red dyewood are boiled for 1 hour in 5.5 lb. water. It is then filtered through a cloth and applied to the article to be stained until the desired color has been obtained. In the meanwhile a solution of 0.175 oz. potash carbonate in 17.5 oz. water has been prepared, and a thin coat of this is applied to the article stained red. But strict attention must be paid not to apply too thick a coat of this solution, or else a dark blue color would be the result.

Red.—(1) The wood is plunged first in a solution of 1 oz. of curd soap in 35 fl. oz. water, or else is rubbed with the solution; then magenta is applied in a state of sufficient dilution to bring out the tone required. All the aniline colors behave very well on wood.

(2) For a red stain, a decoction of ¼ lb. logwood and ½ oz. potash in 1 lb. water is used as the bath, being fixed by a wash of alum water. For scarlet, use 1 oz. cochineal, 6 oz. powdered argol, 4 oz. cream tartar, in 12 oz. tin chloride (scarlet spirits).

(3) Take 1 qt. alcohol, 3 oz. Brazil-wood, ½ oz. dragons' blood, ½ oz. cochineal, 1 oz. saffron. Steep to full strength and strain. It is a beautiful crimson stain for violins, work-boxes, and fancy articles.

(4) Beside the aniline colors, which are, however, much affected by sunlight, cochineal gives a very good scarlet red upon wood. Boil 2 oz. cochineal, previously reduced to a fine powder, in 35 oz. of water for 3 hours, and apply it to the wood. When dry, give it a coating of dilute tin chloride to which is added a little tartaric acid—1 oz. tin chloride and ½ oz. tartaric acid in 35 fl. oz. water. If, instead of water,

the cochineal is boiled in a decoction of bark (2 oz. bark to 35 oz. water), and the tin chloride is used as above, an intense scarlet and all shades of orange may be produced according to the proportions.

(5) Take 1 gal. alcohol, 1½ lb. camwood, ½ lb. red sanders, 1 lb. logwood extract, 2 oz. aqua fortis. When dissolved, it is ready for use. It should be applied in 3 coats over the whole surface. When dry, rub down to a smooth surface, using for the purpose a very fine paper. The graining is done with iron rust, and the shading with asphaltum thinned with spirits of turpentine. When the shading is dry, apply a thin coat of shellac; and when that is dry, rub down with fine paper. The work is then ready for varnishing—a fine rose tint.

(6) Monnier recommends steeping the wood for several hours in a bath of 1200 gr. potassium iodide to the quart of water, and then immersing it in a bath of 375 gr. corrosive sublimate, when it will assume a beautiful rose-red color by chemical precipitation. It should subsequently be covered with a glossy varnish. The baths will not need renewal for a long time.

(7) 2.2 lb. finely powdered Lima red dyewood and 2.1 oz. potash carbonate are put in a glass bottle and digested in 5.5 lb. water for 8 days in a warm place; the bottle should be frequently shaken. It is then filtered through a cloth; the fluid is heated, and applied to the article to be stained until the latter acquires a beautiful color. If it is desired to brighten the color, a solution of 2.1 oz. alum, free from iron, in 2.2 lb. water is applied to the article while it is still wet. The last solution can be prepared by heat; when it has been accomplished, it is filtered. As soon as the stains have become dry, they should be rubbed

with a rag moistened with linseed oil, after which the varnish may be applied.

Satinwood.—Take 1 qt. alcohol, 3 oz. ground turmeric, 1½ oz. powdered gamboge. When steeped to its full strength, strain through fine muslin. It is then ready for use. Apply with a piece of fine sponge, giving the work 2 coats. When dry, sandpaper down very fine. It is then ready for polish or varnish, and is a good imitation of satinwood.

Violet.—The wood is treated in a bath made up with 4¼ oz. olive oil, the same weight of soda-ash, and 2½ pints boiling water, and it is then dyed with magenta to which a corresponding quantity of tin crystals has been added.

Walnut.—Deal and other common woods are stained to imitate polished walnut in various ways. (1) One method is, after careful rubbing with glass-paper, to go over the surface with a preparation of Cassel brown boiled in a lye of soft-soap and soda. After drying, the surface is rubbed over with pumice and oil, and polished with shellac. The Cassel brown will not take equally well on all kinds of wood, so that if not laid on thick it sometimes comes off under the subsequent pumicing; whilst on the other hand this same thickness conceals, more or less, the grain on the wood beneath, giving it the appearance of having been painted.

(2) Others use instead a decoction of green walnut-shells, dried and boiled in the same lye, or in soft water to which soda has been added. The decoction of walnut-shells is apt to come off on the clothes as a yellowish, adhesive substance.

(3) Others, again, employ catechu and potash chromate in equal parts, boiled separately and afterwards mixed. The mixture of catechu and potash

chromate leaves a reddish-brown deposit on the surface of the wood, very unlike real walnut.

(4) The following is said to be a very superior method for staining any kind of wood in imitation of walnut, while it is also cheap and simple in its manipulation. The wood, previously thoroughly dried and warmed, is coated once or twice with a stain composed of 1 oz. extract of walnut peel dissolved in 6 oz. soft water by heating it to boiling, and stirring. The wood thus treated, when half dry, is brushed with a solution of 1 oz. potash bichromate in 5 oz. boiling water, and is then allowed to dry thoroughly, and is to be rubbed and polished as usual. Red beech and alder, under this treatment, assume a most deceptive resemblance to American walnut. The color is fixed in the wood to a depth of one or two lines.

(5) Mix dragons' blood and lampblack in methylated spirits till you get the color required, and rub it well into the grain of the wood.

(6) LIGHT WALNUT.—Dissolve 1 part potassium permanganate in 30 of pure water, and apply twice in succession; after an interval of 5 minutes, wash with clean water, and when dry, oil and polish.

(7) DARK WALNUT.—Same as for light walnut, but after the washing with water the dark veins are made more prominent with a solution of iron acetate.

(8) In the winter season get some privet berries (black), which grow in most gardens, and put 2 oz. in ½ pint solution of liquid ammonia. This, applied to pine, varnished or polished, cannot be detected from real walnut itself.

(9) Take 1 gal. very thin sized shellac; add 1 lb. dry burnt umber, 1 lb. dry burnt sienna, and ¼ lb. lampblack. Put these articles into a jug and shake

frequently until they are mixed. Apply one coat with a brush. When the work is dry, rub down with fine paper, and apply one coat of shellac or cheap varnish. It will then be a good imitation of solid walnut, and will be adapted for the back boards of mirror-frames, for the back and inside of casework, and for similar work.

(10) Take 1 gal. strong vinegar, 1 lb. dry burnt umber, ½ lb. fine rose pink, ½ lb. dry burnt Vandyke brown. Put into a jug and mix well; let the mixture stand one day, and it will then be ready for use. Apply this stain to the sap with a piece of fine sponge; it will dry in ½ hour. The whole piece is then ready for the filling process. When the work is completed, the stained part cannot be detected even by those who have performed the job. By means of this recipe, wood of poor quality and mostly of sap can be used with good effect.

(11) DARKENING WALNUT.—Slaked lime, 1 to 4 of water, will do for some kinds of walnut; a weak solution of iron sulphate for others; and yet again for other kinds a weak solution of pearlash. Try each on the wood, and choose the one you like best.

(12) To give to walnut a dark color resembling rosewood, Hirschberg uses a solution of 0.17 oz. potash bichromate in 1.05 oz. water. This solution is applied to the walnut with a sponge, and the wood is then pumiced and polished.

(13) By a simple staining, furniture of pine or birch wood can be easily made to appear as if it had been veneered with walnut veneer. For this a solution of 3.15 oz. potash manganate, and 3.15 oz. manganese sulphate in 5.25 qt. hot water, is made. This solution is applied to the wood with a brush, and must be

repeated several times. The potash manganate is decomposed when it comes in contact with the woody fiber, and thus a beautiful and very durable walnut color is obtained. If small wooden articles are to be stained in this manner, a very diluted bath is prepared; the articles are dipped into it, and kept there 1 to 9 minutes, according as the color is desired lighter or darker.

Yellow.—(1) Mordant with red liquor, and dye with bark liquor and turmeric.

(2) Turmeric dissolved in wood naphtha.

(3) Aqua regia (nitro-muriatic acid), diluted in 3 parts water, is a much-used though rather destructive yellow stain.

(4) Nitric acid gives a fine permanent yellow, which is converted into dark brown by subsequent application of tincture of iodine.

(5) Wash over with a hot concentrated solution of picric acid, and when dry, polish the wood.

(6) ORANGE-YELLOW TONE TO OAK WOOD.—According to Niedling, a beautiful orange-yellow tone, much admired in a chest at the Vienna Exhibition, may be imparted to oak wood by rubbing it in a warm room with a certain mixture until it acquires a dull polish, and then coating it after an hour with thin polish, and repeating the coating of polish to improve the depth and brilliancy of the tone. The ingredients for the rubbing mixture are about 3 oz. tallow, ¾ oz. wax, and 1 pint oil of turpentine, mixed by heating together and stirring.

(7) 0.5 oz. nitric acid (aqua fortis) is compounded with 1.57 oz. rain-water, and the article to be stained is brushed over with this. Undiluted nitric acid gives a brownish-yellow color.

(8) 2.1 oz. finely powdered turmeric are digested for several days in 17.5 oz. alcohol 80 per cent strong, and then strained through a cloth. This solution is applied to the articles to be stained. When they have become entirely dry, they are burnished and varnished.

(9) 1.57 oz. potash carbonate are dissolved in 4.2 oz. rain-water. This solution is poured over 0.52 oz. annotto, and this mixture is allowed to stand for 3 days in a warm place, being frequently shaken in the meanwhile. It is then filtered, and 0.175 oz. spirit of sal-ammoniac is added to it. The stain is now ready, and the articles to be stained will acquire a very beautiful bright yellow color by placing them in it.

(10) BRIGHT GOLDEN YELLOW. — 0.52 oz. finely powdered madder is digested for 12 hours with 2.1 oz. diluted sulphuric acid, and then filtered through a cloth. The articles to be stained are allowed to remain in this fluid 3 to 4 days, when they will be stained through.

Most of the foregoing is taken from English, French, and German sources, but the following are methods usually adopted in American practice; but it is just as well that the finisher should be conversant, to some extent, with the practice of other nationalities as well as that of his own.

Stains may be classified as follows: Those made with oil and color, those made with spirits and color, those made with water and coloring matter other than anilines, and those made with water and anilines. Generally, spirit stains are made with anilines.

Water stains are those in which the coloring matter is mixed with water. A good mahogany water stain is made as follows: Fustic chips, 4 oz.; madder root, ½ lb.; water about 4 qt. This should boil for several hours and should be applied boiling hot, after being well strained.

For walnut: Vandyke brown, 1 lb.; strong lye, ½ pt.; water, 3 qt. This should boil down to about three pints, and should be applied to the wood when cold with a sponge or pad.

A good cherry stain can be made by mixing a pound of Spanish annotto, an ounce of strong lye, and water. Boil forty minutes. If not deep enough in color, boil more, and add a little gamboge to darken it.

There is very little difference between cherry and mahogany stains; the latter are somewhat darker, but may be reduced to answer.

Aniline stains are easily made, and are usually very even and free from cloudy spots when properly laid on. All or any of anilines will dissolve in water, oil, or alcohol, but will dissolve much quicker in warm liquid than in cold. Time will therefore be saved by having the medium hot.

One example of making a water stain from anilines will answer for the whole group. If for mahogany, use Bismarck brown, 1 oz.; add 3 qt. water boiling hot; stir until the brown is all dissolved. When cool it is ready to apply.

Many excellent stains for pine may be obtained by using the ordinary graining colors, Vandyke brown, raw and burnt sienna, ultramarine blue, etc., applied with a brush, without previous preparation, and then wiped off with a cloth—a method that brings out clearly the grain or marks of the wood, which in pitch pine, now being extensively used for fittings, are often extremely beautiful. A better method for general work, French polish being ordinarily too expensive, is, where dark oak or mahogany stains are not wanted,

light varnishes, of which two coats are to be applied.
The glue size with which the work is first coated, in
order to fill up the pores of the wood, should not be
too thick, as in that case it is liable to crack.

Logwood, lime, brown soft soap, dyed oil, sulphate
of iron, nitrate of silver exposed to the sun's rays,
carbonate of soda, bichromate and permanganate of
potash, and other alkaline preparations, are used for
darkening the wood; the last three are specially
recommended. The solution is applied by dissolving
one ounce of the alkali in two gills of boiling water,
diluted to the required tone. The surface is saturated
with a sponge or flannel, and immediately dried with
soft rags. The carbonate is used for dark woods. Oil
tinged with rose madder may be applied to hardwoods
like birch, and a red oil is prepared from soaked
alkanet root in linseed oil. The grain of yellow pine
can be brought out by two or three coats of japan,
much diluted with turpentine, and afterwards oiled
and rubbed. To give mahogany the appearance of
age, lime water used before oiling is a good plan.
In staining wood, the best and most transparent effect
is obtained by repeated light coats of the same. For
oak stain a strong solution of oxalic acid is employed;
for mahogany, dilute nitric acid. A primary coat or a
coat of wood fillers is advantageous. For mahogany
stains, the following are given: two ounces of dragons'
blood dissolved in one quart of rectified spirits of
wine, well shaken, or raw sienna in beer, with burnt
sienna to give the required tone; for darker stains boil
half-pound of madder and two ounces of logwood
chips in one gallon of water, and brush the decoction
while hot over the wood; when dry, paint with a
solution of two ounces of potash in one quart of water.

A solution of permanganate of potash forms a rapid and excellent brown stain.

Oak or ash may be stained brown by using linseed oil and benzine half and half, and burnt umber or Vandyke brown incorporated with this. Maple can be stained green-gray by using copperas in water; oak will also be changed to a dark green blue through the same agency, the effect on ash being various shades of olive green. Ammonia applied to oak produces the bronze olive tint now used so much by architects.

Wash any compact wood with a boiling decoction of logwood three or four times, allowing it to dry between each application. Then wash it with a solution of acetate of iron, which is made by dissolving iron filings in vinegar. This stain is very black, and penetrates to a considerable depth into the wood, so that ordinary scratching or chipping does not show the original color.

A wash of one part of nitric acid in ten parts of water will, if well done, impart a stain resembling mahogany to pine wood which does not contain much resin. When the wood is thoroughly dry, shellac varnish will impart a fine finish to the surface. A glaze of carmine or lake will produce a rosewood finish. A turpentine extract of alkanet root produces a beautiful stain which admits of French polishing. Asphaltum, thinned with turpentine, makes an excellent mahogany color on new wood.

When describing the treatment of different kinds of wood, which will follow, I will have more to say regarding the method of staining.

VARNISHING AND POLISHING

French polishing was at one time the only method of polishing permitted to be employed on work of the first class, because of its permanency and uniform appearance, and the man who understood the process, and could mix the materials used, could always demand good pay and steady employment. Now, however, there are a number of methods and varnishes that are almost as good as the method and materials used in French polishing, and which can be applied at one-eighth the cost, and which when done look just about as well, though perhaps they will not be permanent. So, while French polishing still survives to some extent, and is likely to be practiced for many years to come, a description of the method and materials is quite necessary in a book of this kind; therefore no excuse is necessary to account for its appearance in these pages.

Varnishing, on which depends to a certain extent the beauty of the work, to be of a durable character should be done in a temperature of 65° F. or more, from the beginning of the work to the finishing of the same, day and night included. A higher temperature, if not over 125° F., will not harm fine varnishes; in fact, will turn out nicer work than in a lower temperature, and will enable a manufacturer to turn out work in a much shorter time. In a temperature of but 65° to 75° F. from four to six days between coatings is advisable, as this will give good and unfailing results. In 125° F. the same good results can be obtained in one-half the time. Varnishing departments, outside the filling and rubbing rooms, should be kept absolutely clean. The filling and rubbing rooms should be kept as clean as possible.

After cabinet-work has received one-half the varnish coatings and the varnish is perfectly dry, rub the surface with pumice-stone and water—use a piece of felt—to a smooth, even surface. Allow the work to stand 24 hours, and then begin the application of the last half of varnish coatings, giving the same time between coatings. After all the coatings are perfectly dry, go through the same rubbing process. A perfect, smooth surface for polishing will be the result. Let the work stand for 24 hours after this rubbing, then start polishing by moistening a fine piece of cloth with water, dipped in powdered rotten stone, thus moistening the same also, and begin to rub the surface of the work with a steady hand and evenly, in order to remove with this fine rotten stone the fine scratches, if any, which are generally caused by the rubbing of the pumice-stone. This accomplished, continue the rubbing with the palm of the hand instead of the cloth, using moist rotten stone, and rub the work until the fine polish required is obtained. The rotten stone then generally falls off the hand and you work in a dry dust. Wash the surface clean with water, using a fine sponge and chamois. Allow the polished work to stand 24 hours, then oil the same off with a light oil and a very soft rag or cotton bat. Take another fine rag or cotton bat and remove all the oil by rubbing or wiping the same gently, but absolutely clean, off the polished surface. To be sure this is accomplished, moisten the cloth or cotton bat with alcohol. The polish, if everything is done correctly, will then be finished.

Care should be taken that too much varnish is not put on the work. For good finishing only a minimum of material should be used if the grain of the wood is wished to be seen, for the less varnish used, providing a good polish is obtained, will bring out the details of the wood better than a dozen coats would. If one coat of varnish is not enough, two will be, and it is not good practice to employ more, nor is it good workmanship.

A room that is dark or damp will spoil the varnish, and a room that is excessively warm will keep the varnish soft. Barrels of varnish should not be stored in a very warm room, as the heat will open the joints in the barrels, and the varnish, which will be quite thin owing to the heat, will leak out. In summer, varnish should be stored in a basement where it is cool.

Turning white is caused by the action of water or dampness. The better the grade of varnish and the more elastic it is, the less liable it is to be affected by moisture. In applying two coats of varnish, neither should be heavy, more especially the first coat. If it is, it will not generally get thoroughly dry at the bottom before the second one is applied, and the result will be disastrous, as it will cause the last coat to crack, wrinkle, or sag, as it is called. Another bad result of not giving varnish time to dry will be noticed by the last coat deadening or sinking away. This is caused by the undercoat not having been allowed sufficient time to dry, resulting in the finishing coat becoming absorbed while in the course of hardening.

The varnisher must be careful that there is no oil on the surface to be varnished, as it will cause pitting, or little hollows in the varnish. When varnish is so thick it does not spread well, it may be thinned by adding a little turpentine, but care must be taken that it is thoroughly incorporated in the varnish, as, if it is not, this will also cause pitting. A long-haired, pliant brush is best with which to apply varnish. It should be spread with long, steady strokes, drawn in a perfectly straight line lengthwise the grain of the wood. Oil japan or liquid dryer should never be added to good varnish. Each coat of varnish should be given three or four days to harden before another coat is added or before the last coat is rubbed, if a rubbed surface is required.

One thing should always be considered before varnishing begins, and that is, that a good coat of colorless shellac should be laid on the work just as soon as the filler is dry and properly rubbed down. This will hermetically seal up all the pores and prevent the varnish from sinking or showing hollow spots on the work.

I show the style of brush which is generally used for applying shellac, in Fig. 107; it is taken from a dealer's catalogue, and is oval in section and metal-bound.

An excellent shellac varnish is made by dissolving 3½ lb. of orange shellac in 1 gal. of grain or wood alcohol. Let stand in a warm place until the shellac is thoroughly dissolved. If shaken once in a while, the shellac will dissolve quicker.

For making white shellac varnish, take 3¼ lb. of white shellac and mix with alcohol, grain or wood, the same as before. These proportions are offered only as a sort of guide, but they may be varied a little as experience will show.

Fig. 107

All varnish should be laid on with the grain if possible, or there will likely be sagging along the brush marks.

Where varnish is flowed the operator should not be "stingy" with his material or his work will not be good. There is a tendency always with the beginner to "lay off" his varnish too thin, or to "skin" it as it is termed. It is always better to err on the side of using too much, than too little varnish, if good work is desired. Too much varnish, however, should be avoided, as the work will be apt to crack and become disfigured in time. The brushes shown in Figs. 108 and 109 are among the best for flowing purposes.

Generally, unless on the very hardest of woods, two coats of varnish are necessary before the process of rubbing can be commenced, as a good surface cannot be obtained unless there is a good ground to work on.

A few hints as to "rubbing" may not be out of place. First, see that the varnish is dry and hard. If an impression can be made on it with a finger nail, it is too soft to rub; let it stand awhile.

Never rub across the grain if it can be avoided.

Always rub with the grain, lightening the stroke at the ends.

Rub lightly at first and avoid making bare spots.

FIG. 108

Use fine pumice-stone at first, and all the time, until you get accustomed to the work.

Use a pad of felt about 3½x4½ inches for a rubber.

Rub with either water or with rubbing oil—the latter preferred—or use linseed oil.

Sprinkle the pumice-stone powder on the work, dip the rubber in the oil, then rub lightly.

Clean off the work with damp, clean sawdust and a brush, or with soft cotton wadding.

For mouldings, make reverse shapes, cover with felt, and proceed as with the flat surfaces.

After rubbing, the work is ready for polishing. This is done by making a mixture of half sweet oil and half alcohol. Then make a large wad of nice clean cotton and rub the work with a circular motion until the proper polish is attained. This requires some practice, but is easily acquired.

Another and a better method is to put on an extra coat of flowing varnish, then rub down with fine pumice-stone and clean carefully. After this, rub down with a lump of faced-up pumice, or with chamois leather and powdered rotten stone. Let the rotten stone become dry on the surface, then wipe off with the palm of the hand, which rubs up the work, with a rotary motion. A piece of fine linen or silk should be handy, on which the hand should be cleaned from time to time.

FIG. 109

If a dead finish is required, do not polish after rubbing over with powdered pumice-stone and oil.

Of course, the operator must always bear in mind, during the several operations, that care and cleanliness are two necessary factors, and without them, there can be no such thing as satisfactory results. A slovenly workman should never attempt to polish, for no good can come from it. In all my experience, which covers many years, I have never known of a careless or slovenly man making good work in this particular department.

French Polishing.—Much has been said and written on this subject, and many comparisons have been made between this and other methods of polishing, but the writer, from his architectural experience and general observation, does not hesitate for a moment in saying that "French polished" work is much superior to any or all other polished work, so far as durability and effect is concerned. True, other methods are—many of them—much more economical and easier acquired, but they do not have the staying powers that the good old-fashioned French polishing has.

Although, as before stated, much has been written on the subject, there is a certain sameness in the descriptions, and I confess I do not see how it could be otherwise, as the subject is one that can have only one side to it; hence the sameness of description.

I will not attempt to write a description, as it would be no improvement on former ones; so I offer the following, which is taken from an English source, that has been Americanized to some extent.

In French polishing, the wood has to be prepared and various minor details attended to before this can be done. For example, the pores of open-grained wood must be stopped or, as the process is generally called, filled, to get a smooth surface and to prevent excessive absorption of the liquid polish. Then the appearance of some woods is improved and enriched by oiling them before applying the polish. This oiling, to a certain extent, darkens and mellows them, and brings up the figure.

The temperature and atmosphere of the place in which French polishing is done are of considerable importance. Work cannot be done properly in a cold or damp room, as then the polish will get chilled, and, as it sets on the wood, becomes opaque and cloudy. To avoid this the polisher should work in a warm room. The temperature for a living room, about 70°, is about that suited for polishing. In warm summer weather a fire is not necessary, but in winter it is. If the polisher notices that his polish chills, he must increase the heat of his room. If a moderate amount of warmth be brought near the surface as soon as any chill is observed, it will probably disappear. A small article may be taken to the fire, but with large work this course would hardly be convenient. In such cases a good plan is to hold something warm a short distance from the chilled surface, but on no account must it touch, nor must the heat be great enough to scorch the polish. A common plan, but not altogether a good one, is to hold a piece of burning paper near the chill. An ordinary flatiron is very useful for small chilled patches. When the article is cold or damp, chill is more likely to occur. It is, therefore, always necessary to make sure after a stain has been used that the wood has become thoroughly dry.

Not less important is the employment of suitable materials, both in the polish and in the tools of the polisher's art. These latter consist almost entirely of wadding or cotton wool and soft linen or cotton rags, from which the rubbers to apply the polish are made;

and a few bottles are wanted to hold the **various** polishes, stains, and their components.

The pad with which French polish is applied is called the rubber. Without it the French polisher can do little in actual polishing, although he may not require it in the preliminary operations of oiling and staining. However simple in itself the rubber may be, it should be properly and carefully made of suitable materials; otherwise good work cannot be done with it. Those who have seen polishers at work may be inclined to infer that no great care is necessary, for a dirty rag covering an equally uninviting lump of wadding is usually seen. Examination will show the rubber to be more carefully made than might have been expected, and the expert polisher would probably prefer it to a nice clean-looking rubber such as a novice would choose. Nevertheless, a dirty rubber is not wanted, for dirt is fatal to first-class work; hence the polisher should keep his rubbers scrupulously clean. They will naturally get stained and discolored with the polish, but that is a very different matter from being dirty. Old rubbers are preferable to **new** ones, provided they have been properly taken care of and not allowed to get hard.

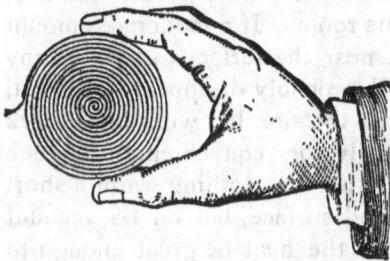

FIG. 110

For flat surfaces or fretwork a wad may be prepared by using a strip of torn woolen cloth from 1 in. to 2 in. wide. Cloth with a cut edge is not recommended for this purpose, as it is too harsh. Roll the strip very tightly into a wad about 1 in., 2 in., or 3 in. diameter, according to the size of the work, and tie tightly round with fine twine (Fig. 110). This will give as nearly as possible a rubber resembling Fig. 111. This wad is put into a double thickness of linen cloth, and the ends are gathered up like the ends of a pudding-cloth; they are not tied, but are grasped as a hand-piece while being used. This form of rubber would, however, be use-

Fig. 111

less for bodying up mouldings, beads, quirks, moulded handrails, newel posts, etc., and when polishing large mahogany doors or other framed furniture it would be impossible to get well into the corners of sunk panels, as in Fig. 112.

A well-shaped, soft, pliable rubber, with its rag covering free from creases, is to a practical French polisher equivalent to a sharp, finely set smoothing plane in the hands of a cabinet-maker. With such a rubber, made of wadding, one is enabled to get into corners, round turned work, and up to

Fig. 112

the edges of mouldings in a manner impossible with a hard, round rubber. To make it, take a sheet of wadding—this is 9 in. wide—and tear off a piece 6 in. long; this will form a conveniently-sized rubber, suitable for most work; but for small work use one of smaller size. Double the wadding, making it 6 in. by 4½ in.

Squeeze this in the hand, keeping the skin unbroken, till in shape it nearly resembles half a bear. The rubber should then be charged with polish, and covered with a piece of soft, clean rag. In folding the rag, twist it on the upper side of the rubber. Each additional twist will bring it to a sharper point and cause the polish to ooze through its surface. It is not necessary to put a rag covering on the rubbers first used. The rubber must be kept free from creases, otherwise it will cause the surface to which it is applied to be stringy or full of ridges.

Though rags have been mentioned as suitable for the outer covering or casing of the pad, some care in their selection is necessary. A piece with a seam across it would never do for a rubber, and anything which would tend to scratch the film of polish as it is being laid or worked on in the final operation of spiriting off must be carefully avoided. They must be either cotton or linen, and ought to be perfectly soft and fine or, at any rate, free from knots or lumps. Some polishers advise the exclusive use of linen, but this is a needless restriction. It may be mentioned that new material may be used as well as rags. To render this suitable, all traces of the sizing and stiffening with which it may have been finished must be removed by a thorough washing.

Any material to be used for a polishing rubber must be thoroughly well dried. Indeed, the necessity of avoiding damp cannot be too much insisted on. With regard to the substance of the rubber, white wadding is the best to use, and this is readily obtainable from any upholsterer or chemist. It may be purer if got from the latter, but it is certainly much dearer than any reasonable upholsterer would charge for something equally suitable. Wadding bought from an upholsterer has a thin skin on one or both sides, according to whether it has been split or not. This skin must be removed, leaving nothing but the soft cotton. For a few cents enough ought to be obtainable to last a considerable time. It may be useful to know that, if it is clean, the raw material used in cotton manufacturing districts will do as well as the finest wadding. Cotton flock, used by upholsterers as a stuffing for mattresses, chairs, etc., is not suitable for polishing, except perhaps for the coarsest work. Even for this it should not be used if anything better is available. Rubbers composed entirely of flannel are occasionally recommended for special kinds of work. It is doubtful if there is any advantage in using flannel, except for large, flat surfaces, which can be got over more quickly with a large rubber than with a small one. The novice is advised to use the wadding rubber, and to become an expert polisher with it before experimenting with anything else.

The size of the rubber will, to some extent, depend on the nature of the work, but that above given may be regarded as generally suitable. A very large rubber is not advisable at first, and the polisher, as he gains experience, must be guided by circumstances. In handling it, moreover, the polisher will be equally guided; a rubber of moderate dimensions is usually held by the tips of the thumb and fingers, but the polisher will probably find a large rubber can be more conveniently used by holding it in the palm of the hand.

The rubber must be charged with polish for use, and some care will have to be exercised in doing this. The covering of the rubber is opened so that a little polish can be dropped on the wadding. A convenient way of doing this is to have the polish in a bottle, the cork of which has a channel or notch cut in it to allow only a few drops to escape at a time. Some polishers dip a portion of the rubber into the polish, but the other method is more generally adopted. It must not be saturated; only enough polish to moisten the wadding must be used, or what will appear through the rag covering when pressed. The rubber having been thus charged, gather up the edges of rag as before directed. Then, to distribute the polish equally, press the rubber moderately firmly into the palm of the other hand. The rubber ought now to be ready for application to the wood, which may be assumed to have been properly prepared to receive its first coating of polish.

FIG. 113

At this stage the principal thing is to get a good body of polish evenly spread on the wood. How this may best be done depends on circumstances; but if the desired result is obtained, the precise method is of secondary consequence. Let it be assumed that the work to be done is a small, flat surface. With moderate pressure on the rubber, quickly wipe over the entire surface, first with the grain of the wood, then across it. Then, without delay, go over it more minutely, the motion generally adopted for the rubber being shown in the illustrations (Figs. 113 and 114). At first the pressure should be gentle, but it should be increased as the polish gets worked in and the rubber drier, though at no time must the rubbing decline to scrubbing.

While the rubber is in contact with the wood it must be kept constantly in motion. An important point is not to allow the rubber to remain stationary on the woodwork during temporary absence or at the end of the day's work. As the rubber gets dry it must be recharged with polish, but let the novice beware of using this in excess.

Old rubbers are better than new, so when done with they should be kept in an airtight receptacle, such as a tin canister or a biscuit box. When any job is finished, do not throw the rubber away under the impression t h a t a rubber once laid aside b e c o m e s useless. T h i s o c c u r s only when it is left exposed to the air, because then it hardens throughout. I f , a s stated, the rubbers are kept from the air, only the outside covering will get dry and

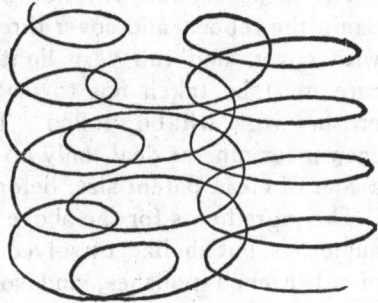

FIG. 114

hard, and this can be readily softened by rubbing it in raw alcohol.

Every time the rubber is wetted with polish it should be pressed in the palm of the left hand, which will equalize the polish. After the cover is put over, which should be some clean old cotton or print rags, the tip of the finger should be dipped in linseed oil and applied to rubber cover—just enough to keep it from sticking. As soon as the cover has a shiny appearance, it should be removed to a fresh place. As soon as the work has got a good body of polish on, it should be set away for at least ten hours, to allow for the polish to sink, which always takes place.

Before commencing to polish again, the work should be very carefully rubbed over with the finest glass-paper obtainable, taking care not to cut through the skin; then proceed as before. Be sure never to let the rubber stop in one place for an instant, as it will surely take off the polish to the bare wood and spoil the job. After the work has sufficient polish on, it should be allowed to stand three or four hours before spiriting off.

The same rubber will do; only use spirits. Just damp the rubber, and cover three or four times double with cover, and rub very lightly over the work; but care must be taken not to make the rubber too wet, or the work will be spoiled. The same process will answer for pine or deal, only no filling is required, but a coat of clean patent size, before applying the polish.

The ingredients for the above kind of work are quite numerous, but shellac, dissolved in alcohol, is the basis of all French polishes, and some finishers use thin shellac varnish without other admixture, slightly moistening the rubber with linseed oil to prevent stickiness and make it work smoothly. There is a great variety of admixtures and diversity in the proportion of ingredients, but the differences are not material. I subjoin a number of recipes.

First and Best.—To one pint of spirits of wine add a quarter of an ounce of gum-copal, a quarter of an ounce of gum arabic, and one ounce of shellac.

Let the gums be well bruised, and sifted through a piece of muslin. Put the spirits and the gums together in a vessel that can be closely corked; place them near a warm stove, and frequently shake them. In two or three days they will be dissolved. Strain the mixture through a piece of muslin, and keep it tight corked for use.

Next.—Take one ounce each of mastic, sandarac, seed lac, shellac, gum lac, and gum arabic; reduce them to powder; and add a quarter of an ounce of virgin wax; put the whole into a bottle, with one quart of rectified spirits of wine; let it stand twelve hours, and it will be fit for use.

Another.—Put into a glass bottle one ounce of gum lac, two drachms of mastic in drops, four drachms of sandarac, three ounces of shellac, and half an ounce of gum dragon; reduce the whole to powder; add to it a piece of camphor the size of a nut, and pour on it eight ounces of rectified spirits of wine. Stop the bottle close, but take care, when the gums are dissolving, that it is not more than half full. Place near a warm stove until dissolved.

Other French Polish Recipes.—One pint naphtha, 3½ oz. orange shellac, ½ oz. elima. Darken with red sanders wood.

To one pint of spirits of wine, add half an ounce of gum shellac, half an ounce of seed lac, and a quarter of an ounce of gum sandarac; submit the whole to a gentle heat, frequently shaking it, till the various gums are dissolved, when it is fit for use.

Shellac, 6 oz.; naphtha, 1 qt.; sandarac, 1 oz.; benzoin, ¾ oz.

Three oz. shellac, ½ oz. gum mastic pulverized, and 1 pt. methylated spirits of wine added. Let it stand till dissolved

Twelve oz. shellac, 2 oz. gum elima, 3 oz. gum copal, 1 gal. spirits of wine; dissolve.

The following must be well mixed and dissolved:

pale shellac, 2¼ lb.; 3 oz. mastic, 3 oz. sandarac, 1 gal. spirits of wine. After the above is dissolved, add 1 pt. copal varnish, 1¼ oz. shellac, ½ oz. gum juniper, ½ oz. benzoin, ½ pt. methylated alcohol.

A Good Polish.—To 1 pt. spirits of wine add, in fine powder, 1 oz. seed lac, 2 dr. gum guaiacum, 2 dr dragons' blood, and 2 dr. gum mastic; expose them, in a vessel stopped close, to a moderate heat for three hours, until you find the gum dissolved; strain the whole into a bottle for use, with a quarter of a gill of the best linseed oil, to be shaken up well with it.

This polish is more particularly intended for dark colored woods—for it is apt to give a tinge to light ones, as satinwood or airwood, etc.—owing to the admixture of the dragons' blood, which gives it a red appearance.

A Polish That Will Stand Water.—Take 1 pt. spirits of wine, 2 oz. gum benzoin, ¼ oz. gum sandarac, and ¼ oz. gum animé; these must be put into a stopped bottle, and placed either in a sand-bath or in hot water till dissolved; then strain the mixture, and, after adding about a quarter of a gill of the best clear poppy oil, shake it well up, and put it by for use.

Prepared Spirits.—This preparation is useful for finishing after any of the foregoing recipes, as it adds to the luster and durability, as well as removes every defect, of the other polishes and it gives the surface a most brilliant appearance.

Half a pint of the very best rectified spirits of wine, 2 dr. shellac, and 2 dr. gum benzoin. Put these ingredients into a bottle, and keep it in a warm place till the gum is all dissolved, shaking it frequently; when cold, add two teaspoonfuls of the best clear white poppy oil; shake them well together, and it is fit for use.

This preparation is used in the same manner as the foregoing polishes; but, in order to remove all dull places, you may increase the pressure in rubbing.

Polish for Turner's Work.—Dissolve 1 oz. sandarac in ½ pt. spirits of wine; shave 1 oz. beeswax, and dissolve it in a sufficient quantity of spirits of turpentine to make it into a paste, add the former mixture to it by degrees; then, with a woolen cloth, apply it to the work while it is in motion in the lathe, and polish it with a soft linen rag; it will appear as if highly varnished.

A French Polish Reviver.—Beat gum acacia and white of two eggs in a mortar until they amalgamate; then add ½ pt. raw linseed oil and best vinegar, 8 oz. methylated spirits of wine, 1 oz. hydrochloric acid and 2 oz. muriate of antimony. They are to be rubbed on the surface of the furniture until dry, and will give a brilliant and lasting polish.

It now remains to explain the several varieties of finishing in use; these are largely derived from the peculiar qualities of the different varnishes used. Polishing varnishes, which are very hard and durable, are so called because their surface can be brought to a high luster by rubbing with the proper materials. Flowing or finishing varnishes contain more oil than polishing varnishes, dry more slowly, and are softer, but their peculiar qualities are brilliancy and durability, fitting them for work requiring a brilliant gloss, such as veneered panels. Rubbing varnishes are those that dry sufficiently hard to admit of being rubbed to a smooth surface. Turpentine varnishes, being the cheapest variety, are employed for cheap work, such as common chairs, bedsteads, etc.

DEAD FINISH is a term applied to the finish produced by the reduction of any of the rubbing varnishes with powdered pumice-stone and raw linseed oil, the surface thus produced being left in the semi-lustrous state by omitting the polishing process. It is now more used than any other for body work, shellac varnish being generally employed because of its adaptation to the requirements of fine cabinet-work, and its properties of quick and hard drying. Copal, animé, and amber varnishes are also used, but are slower drying. The number of coats required depends somewhat upon the quality of the filler, but usually three coats, and sometimes less, are amply sufficient.

Bodying In and Spiriting Off.—The term bodying, applied to the polisher's art, means coating the wood with a thin, evenly distributed layer of the polish. The way in which this is done greatly affects the appearance and the durability of the gloss. When the body is too thin, the gloss subsequently given to it may at first be beautiful, but as the polish sinks or perishes the gloss fades. When the body is too thick the gloss may appear all right, but the work is apt to look treacly, as though varnish had been used; besides, a thick body impairs the pure tone of some woods. The high degree of excellence to which polishing is capable of being brought is seen only on the best cabinet-work. Polish on second-rate furniture or finish is generally in keeping with the inferior quality of the woodwork. The cheap, gaudy furniture which is often seen in shops must not be taken as models of polishing. The price paid for polishing is reduced, with the result that inferior polish is used and less time is spent on the work. Although the best materials and the expenditure of time and labor will not insure good work by unpracticed hands, they are important factors, and it will be wise to use materials of good quality.

To make a good average polish, neither too thick nor too thin, about six ounces of shellac to each pint of methylated or alcohol spirit will be required, but great exactitude in the proportions is not necessary. The proportions may vary according to the fancy of the polisher, and, to some extent, according to the nature of the work he is engaged on. If the polish turns out too thick, it can be thinned by adding more spirit; if too thin, the deficiency can be made up by adding more shellac. A rough-and-ready way of measuring the proportions is to half fill a bottle with the roughly broken shellac, and then fill up with ordinary methylated spirit.

The shellac dissolves gradually, and the process is hastened by an occasional shaking or stirring with a stick. Heat is not necessary; indeed, the preparation of polish by heat is dangerous.

Two kinds of polish are used. One, known as "white polish," is nearly colorless; the other is known as "brown polish," or simply "polish." The latter is always understood if unqualified by the word "white." White polish is made with white or bleached shellac; the other with ordinary orange or reddish-brown shellac.

Either polish may be used on any kind of wood, except where great purity of tint is required. The white is to be preferred for all light woods, such as light oak, ash, sycamore, satin, etc., while the brown may be used on darker; but even on these, white polish is good, with the exception of mahogany, the only ordinary fine wood for which a decided preference might be given to brown polish. Under ordinary circumstances, however, either polish may be used indiscriminately. The point as to brown or white polish for dark wood belongs to the higher branches of the polisher's art.

It will be seen that white polish is the more generally useful of the two, so those who do not care to keep both kinds may confine themselves to it. Through the slightly higher price of the bleached shellac, it costs a little more, but the extra cost is so small that it is hardly worth considering by those who use small quantities. Those who use polish in large quantities can have both kinds.

Polish bought ready-made may be equal to that made at home from the recipe given, for there is nothing to prevent manufacturers using the same ingredients, and many of them do. Still, from the impossibility of knowing the ingredients in ready-made polish, there is some risk attending its use. From the price at which some polish is sold, it is fair to suppose that something cheaper than spirit or shellac has been used; and though good polish may be bought, it is better for the user to prepare his own, which can be depended on. Bought polish may be thoroughly good in every way—brilliant, clear, and durable—but those who are best able to judge generally prefer to make their own polish to do the best class of work. Prejudice may account for this preference.

Manufacturers of polish assert that, in addition to shellac, certain gums or resins improve the quality of the polish, when used with knowledge and discretion. For instance, one gum may give increased elasticity, while another may harden the film; but for a good all-round polish, which can be relied on, many polishers assert that there is nothing to surpass a simple solution of shellac and alcohol. A few approved formulæ for polishes have been given, so that those who feel inclined to do so may experiment for themselves. Shellac is the principal ingredient in nearly all. Those persons who cannot polish with shellac and spirit alone will not be able to do any better with the more complicated mixtures; therefore, no one should remain under the impression that he will do better if he works with another kind of polish.

Enough having now been said about the material, we may proceed to the using of it for bodying. In the first place, the wood must be prepared by filling of one kind or another, and rubbed down smoothly with fine or worn glass-paper, in order to make it fit to receive the polish, for a high degree of finish cannot be got on a rough surface. The rubber, which consists of cotton wadding with a soft rag cover, with which the polish is applied, has been sufficiently treated on, so that nothing more need be said about it. Work, rubber, polish, and a little raw linseed oil being ready, bodying in may be proceeded with in the following way:

Moisten the wadding with polish; put the rag cover on carefully, so that it is without folds or wrinkles. Dab the rubber into the palm of the left hand to distribute the polish evenly, and cause it to moisten the rag at the bottom properly. Supposing the work is a panel or flat surface, the following will be found a good method of treating it, and it is one that is followed more or less closely by experienced polishers.

Rub briskly across the grain to get the surface covered with polish; then by a series of circular movements, as shown by the lines in Figs. 113 and

114, go over the whole of the work. A moderate
pressure should be applied, which should be increased
gradually as the rubber dries, but the movement
should at no time degenerate into mere scrubbing.
In order that the rubber may work smoothly without
sticking, a little raw linseed oil should be applied on
the face of the rubber. The less of this used the
better, and if it can be dispensed with altogether no
harm will be done. To make the rubber work smoothly
a very little will suffice; the tip of a finger, moistened
with oil, and touched on the face of the rubber, is all
that is required. The rubber must not be dipped in
the oil, nor must the oil be dropped on it from a
bottle, for by these means more oil would be applied
than is necessary, and this would be fatal to good
work.

The only recognized oil used in French polishing is
raw linseed. This may be worked over the natural
woods in the first place, to give them that peculiar
tone that cannot be gained by other means; otherwise
the less oil used the better for the durability of the
work. Bear in mind that it forms no part of polish in
itself, being used only to enable us to work the gums
easily; thus, without its aid the polish rubber would be
apt to stick or drag, thus breaking up the surface
instead of leveling it. On any surface in which spirit
varnish forms a part this will be particularly noticeable; and in any case, it is next to impossible to get
that beautiful level surface gained by spiriting out
unless a little oil is used.

As the rubber dries, more polish must be applied to
it, as was done in the first instance, with more oil as
required. A small quantity of polish goes a long way,
and the novice must carefully avoid making the

rubber wet. It should be no more than fairly moist.

Many a beginner, noticing how tedious the work is with a dry rubber, may think that if he used more polish the desired result would be more quickly attained. If the object were merely to get the wood coated, this might be the case; but the result of using too much polish would be that the shellac left by the quick evaporation of the spirit would be ridgy and irregular, instead of in a fine, even coating or body. Anything approaching a flow of polish from the rubber must be avoided. When the rubber is not sufficiently charged with polish, the labor of bodying up will be unduly protracted, or may be rendered impossible if no polish can be rubbed onto the wood.

The first bodying-in process should be continued till it seems that the wood absorbs no more of the polish. There will be a perceptible gloss on its surface, but it will be streaky, and the rubber-marks will show very distinctly. All these marks will be removed later on. It may be thought that, if the polish is too thick or too thin, the result will be very much the same as if the rubber were too wet or too dry. The principal objection to having the polish too thin is that it will take more time in working a good body on the wood. It will, however, be better to risk this rather than to have the polish too thick. An experienced polisher would soon detect fault in either direction by the way in which the polish works, but the novice must be on the look-out for irregularities in the shape of lumps or ridges, and, with a little attention, he will have no difficulty in avoiding serious mishaps.

Let the work stand for at least a day, carefully covered up from dust; on examining it the body will be found to have altered in appearance to an extent which will depend upon how much the polish has sunk into the wood. The work must be again bodied up as before, always remembering to use as little oil as possible. Then it will be again laid aside, and the bodying process repeated till the polish no longer seems to sink in, even after the work has lain aside for a few days. When this stage is reached, the bodying may be considered complete, and the work ready for the first polishing operations. Before proceeding to consider these, however, the novice will do well to note the following hints:

The number of times the work will require to be bodied depends on circumstances. Fine, close-grained woods will not require so many as the more open kinds, such as oak, ash, mahogany, etc.; but for the best work, which is intended to be as durable as can be, it need rarely exceed four. An interval of one or more days may elapse between the successive bodies, the chief object of waiting being to let them sink as much as they will. If, after the work has been laid aside for a few days, the polish has not sunk at all, no advantage would be gained by giving it another body It is very seldom that the first body is enough, but often only one body is applied, where either low price or limited time will not allow of more; so those who wish to do polishing need not think the process cannot be hurried.

Still, imperfect bodying is not advisable, as such work will soon want touching up. When work is made merely to sell, one body, and that of the slightest, is sufficient—from the seller's point of view, if not from the buyer's. Between the bodyings, especially after the first and second, the surface of the

work should be rubbed down with fine glass-paper—
not to rub the body off, but just enough to smooth the
surface. It may here be remarked that pumice
powder, used in moderation, is useful for working
down inequalities of surface. Sandpapering has been
recommended as necessary after the first and second
bodies, but the process may be done after any others,
though it should not be required if the work has been
skillfully done. In fact, the final bodying up may be
regarded almost as the beginning of the spiriting off.

Before beginning to work a fresh body on a previous
one, it is as well to wash the surface gently with
lukewarm water, not using too much of it, in order to
remove the grease and allow the rubber to work freely.
The water must be thoroughly dried up before applying the rubber. In moderation the washing can never
do harm, and is, generally, an advantage, though not
absolutely necessary. When a long interval has
elapsed, the washing should never be omitted, as dust
will settle on the work; and it need scarcely be said
that dust should not be rubbed into the polish.

When bodying up, polishers should see that their
hands are clean and free from old polish, which is so
often seen on them. If they are soiled with old polish
or shellac, bits are apt to flake off and destroy the
surface of new work. This may be an appropriate
place to say that any polish which sticks to the hands
may be washed away with hot water and soda, or with
alcohol.

The body should be thin, as it is not so much the
quantity of body on the wood as its quality that is
important; it is also essential that it should be applied
with sufficient intervals between the successive bodies
to allow of sinkage.

Another important matter is to dry the rubbers well by working them on each body till dry, and not to moisten them frequently. By this means the film of shellac is kept thin. Neither a wet nor a dry rubber should on any account be allowed to stand on a surface being polished. The rubber must be kept moving, and should glide gradually on to the work, not be dabbed down on it. In the initial stages of bodying, care in this respect is not so important as later on, when it is absolutely necessary. The same precaution should be used when lifting the rubber from the work.

For the guidance of beginners, it may be stated that if they take care of the edges of the work the rest of the surface will look after itself. The reason is that the edges are apt to be somewhat neglected, and the polish to be less there than elsewhere. The secret of a good, durable polish depends primarily on a good body, and this, in its turn, on sufficient time having been allowed for sinkage.

The final operation in French polishing, by which the gloss is put on the body previously applied, is known as spiriting off. In this operation rubber marks and smears of all kinds are removed, and the beautiful surface, known as French polish, is the result. Bodying is important so far as durability is concerned, but spiriting is more so with regard to finish. If the worker fails in spiriting, his previous efforts will, to a great extent, have been in vain. Disregarding staining, darkening, and other processes, with which a good polisher should be acquainted, the spiriting is perhaps the most severe test of skill in the whole process of polishing; and a man who can manage this part of the work really well may be considered a competent polisher.

The first operation to be described in the process of spiriting off partakes very much of the nature of bodying in. At the beginning it is bodying, and at the end spiriting. The two processes merge one into the other. There is no abrupt break, as between filling and bodying, except for the intermediate stage, although the processes are well defined, both in character and purpose. This intermediate stage is not always practiced, but it is recommended when good work is wanted. Briefly, spiriting off consists in washing the bodied surface with methylated spirit. This being understood, the final bodying up, or first spiriting off, whichever this process may be called, consists in gradually reducing the quantity of polish in the rubber, and supplying its place with spirit. The polish is gradually reduced by the addition of spirit till all the polish has been worked out of the rubber. The rubber may be charged, first with three parts polish and one part spirit; next time equal quantities; the third time three parts spirit and one part polish; and the fourth charging will be with spirit only. It does not follow that these proportions need be strictly observed, nor are they so in practice, but this example illustrates the process. The last rubber will be almost free from polish, and it should be worked till it is dry, or nearly so.

At this stage spiriting proper may begin, and a fresh rubber should be used. It need not be a new one, but it should be one which has been used only for spiriting, and which has no polish on it. It will be better if it has three or four coverings of rag on its face, which can be removed as they dry. If only one cover is used the spirit is apt to evaporate too quickly. The spirit in the rubber has a tendency to partially dissolve the shellac or body on the wood. This it does to a very limited extent, unless the rubber is made too wet, when there is danger of not only spiriting and smoothing the surface, but of actually washing away the body. This mishap must be carefully guarded against. There should be enough spirit to allow the surface of the body to be softened and smoothed, but no more, and the rubbing should be uniform, and not more in one place than in another. There is hardly any likelihood of the novice erring by using too little spirit, so he may be reminded that the less of it there is in the rubber at a time the better. The rubbing should be gentle at first, becoming harder as the spirit dries off, and oil must not be used on the rubber face, for when there is oil either on the rubber or on the work, the polish cannot be brought up.

The chief cause of failure lies in getting the spirit rubber too wet, and so softening and tearing up the gums. Many meet with success by dispensing with it, using instead a swab of clean, soft rag, fairly damp (not wet) with alcohol.

If the spiriting-off process is being done correctly the gloss will soon begin to appear, and when it seems approaching a finished condition, the rubber ought to be moved only in the direction of the grain, and not across it nor with circular motion. The final touches should be given with the soft rubber rag alone, care being taken not to scratch the surface, which is now softened by the action of the spirit. The surface will gradually harden, but for a time it should be handled with care, and nothing be allowed to come in contact with it, or it is very likely to be marked. It should also be protected from dust, for any settling on it may be retained by the polish, the luster of which would certainly suffer.

Hardwood finishers, and perhaps dealers in furniture, who do not keep an experienced polisher, or who may not be able to get their work done out, may be reminded that, circumstances permitting, polished work should be wiped over with a spirit-rubber an hour or two before it is finished, to freshen it up. If the surface is at all soft, neither packing mats, nor anything else likely to injure it, should be allowed to come in contact with it. The polishing on many things sent long journeys is often greatly disfigured from mat and other markings, but they are easily touched up on arrival at their destination.

Some polishers make use of a glaze in order to kill the oil, just before spiriting off, which is made as follows:

Gum-benzoin, 6 oz., dissolved in alcohol spirit 1 pt. Gums other than that mentioned may be substituted in order to cheapen the glaze for commercial purposes, or to suit the fancy of polishers who make up their own solutions. The glaze is used chiefly when leveling down spirit varnish, and for the purpose of imparting a bright finish to mouldings. As practical workers have always at hand rubbers that are specially charged with white polish, brown or red polish, and another rubber for glaze, a clear, bright finish on some kinds of work may be easily gained by passing the glaze rubber along the edges, sunk portions, or even over the whole of a flat surface just before clearing out with the spirit pad. The workman must not, however, suppose that such treatment will entirely kill the oil, if an excessive quantity of oil has been used. Any excess of oil should be cleared off beforehand; or, better still, entirely avoided, if a good, lasting quality of work is desired.

Glazing, though an imitation, has a recognized acceptance among polishers. It is remarkably convenient occasionally, and in some cases possesses an advantage over spiriting, so that it may fairly be classed among the ordinary processes of polishing. When done in moderation, glazing is as useful on furniture carving as the application of spirit varnish. Such work is commonly said, and justly, to be French polished. The real objection to glaze finish is that, though at first the appearance is equal, it is not so durable as the other. Glaze may be said to be even superior to badly spirited finish, and here is the chief claim it has for notice. It is seldom that a casual polisher can manage to do spiriting thoroughly, for the reason that he has not sufficient opportunity of acquiring practice.

Nevertheless, finishing by means of glaze is not so good as the method by spiriting, when the latter is well done, and should be considered as a means of getting the same effect easily and quickly—an imitation, in fact, of the real thing; the difference between the spirit and glaze finishes is that in the one case the effect is produced by friction, in the other by the addition of a thin, fine varnish to the surface of the body of polish. In the former case the polish itself is polished; in the latter it is varnished with a mixture known commonly as glaze, but to which other names are sometimes given.

Among polishers who command a fair price for their work, glaze is of comparatively limited application, and is confined to those parts where the spirit-rubber cannot be conveniently used, or where its use is not necessary. Instances of such may be found in chair-rails and various parts of the frame. These are usually polished, more or less, before the chair is upholstered, or, at any rate, before the outer covering is put on, the finishing being almost necessarily done last of all. The less the chair is handled by the polisher the better, especially if the covering is a delicate one, for there is less risk of injury with one or two wipes over with the glaze rubber than with the more prolonged spiriting.

Glaze may be used with advantage in inlaid work, where the inlay is slightly, though perhaps not intentionally, higher than the surrounding wood. In such a case it is better to resort to glaze than to finish with the spirit rubber. On fretwork, also, glaze may often be used with advantage, and, generally, it is unobjectionable on parts which are not subject to wear and tear. It will stand a moderate amount of handling, but not so much as good, hard spirited-off polish, and the luster is not so durable.

Glaze, under one or other of its different names, may be bought ready-made, but, for reasons similar to those given in connection with French polish, the home-made article is to be recommended. The preparation of glaze is simple, the ingredients being gum-benzoin and methylated spirit. After the benzoin is dissolved, the solution should be strained through muslin to free it from foreign matter. The proportions may vary, but those given for polish do very well, and with the substitution of crushed benzoin for shellac the process of making is exactly the same.

Gum-benzoin differs greatly in quality, but the best should be used by the polisher. Compared with lac it is expensive, so that the saving which is attributed to its use is mainly in time, which is money, at least from a trade point of view. Cheap benzoin is not to be relied on, and in a strange place an experienced polisher would look with suspicion on any offered at a very low price, however satisfactory its appearance. Where material is liable to adulteration, the best way to avoid imposition is to go to a reliable dealer and to pay a fair price.

Glaze, as used by French polishers, can be bought ready-made at most color stores, as patent glaze; it improves with age. To make the genuine article, dissolve 6 or 8 ounces of best gum-benzoin in 1 pint of alcohol spirit. Keep it in a closely stoppered bottle, otherwise the spirit will evaporate quickly.

Glaze may be applied with either rubber, sponge, or brush; in most cases the rubber is most suitable and is most commonly used. It is made in the ordinary way as used for polish, but it must not be applied with pressure. The glaze is painted on rather than rubbed into the work, which must have been previously bodied in. There seems to be an idea that glaze or something put on bare wood will cause a gloss right off; but nothing will do this. A polish can only be got on wood by varnish, or by bodying-in and polish.

When using glaze, the rubber should be made wetter than for polish or spirit; but still there should not be sufficient to drip from it. It should glaze or wet the wood when the rubber is very lightly pressed on it. One or two wipes in the direction of the grain of the wood, with a somewhat quick motion, will put the glaze on. Always let the glaze dry before applying the rubber again to the same place. The coats may be repeated till the gloss is satisfactory, but the film of glaze should never be made a thick one.

If preferred, a sponge may be used exactly as a rubber would be, but it is questionable if there is any advantage gained; it is rather a matter of fancy. When a brush is used, the glaze may be applied as a varnish pure and simple. With a brush a mixture of glaze and French polish, either white or brown, according to the work, in equal quantities, may be used with advantage.

Glaze that is not so satisfactory in appearance as it should be, may sometimes be improved by passing a spirit-rubber lightly over it, though this should be done with great caution, to avoid washing it off. When carefully and skillfully done, there can be little doubt that a glazed surface may be often, if not always, improved by slightly spiriting it.

To glaze a wide surface, see that it is free from dull streaks and ridges and oil, and the rubber soft and free from fluff. Apply the glaze as evenly as possible, going over the surface several times, until the rubber is nearly dry; then, with the smallest quantity of oil and a little spirits, go over the glaze, very lightly at first, varying the direction of the rubber to avoid ridges. A *dull*, even surface may be obtained by adding one-third to one-half of sandarac to the solution of benzoin, and using the rubber only damp—not saturated.

Old French polished work may often be revived by being lightly gone over with glaze after the surface has been washed and cleaned with warm water. This treatment is often considerably better than that commonly adopted with furniture pastes, polishes, creams, and revivers of various kinds.

Wax Polishing.—Although the beauty of most furniture woods is enhanced to the highest degree by French polishing when well done, there are other processes which, though not capable of being brought to such perfection, are much simpler. Among these is wax polishing. This mode of finishing is remarkably easy, both as regards materials and manipulation, and the unskilled novice can manage to wax-polish almost as well as an expert. It is, therefore, a suitable process for the beginner.

Though any wood may be treated by waxing, it is generally confined to oak, especially after this has been darkened by fumigation with ammonia. The appearance of oak so finished is comparatively dull, but it has an attractiveness which French polish does not possess for all eyes.

For antique oak furniture—whether genuine or imitation—wax is the best finish, though varnish is often used. Wax polish, though it may not give the same amount of gloss, is clearer and finer. Varnish clogs the wood, and is apt to give a treacly look to any piece of furniture finished with it.

Mahogany may very appropriately be finished by wax polish, and for many purposes it may be superior to the dulled French polish so often seen. The top of a dining-table is apt to be rendered unsightly from hot plates or dishes injuring the polished surfaces. The heat burns or blisters the hardened shellac of the French polish, and a finish which is not so liable to disfigurement is preferable; this is found in wax polish. Usually, dining-table tops (unless French polished) are simply oil polished. Waxing is, however, less tedious, and at least as suitable for the purpose, and the readiness with which an accidental marking can be obliterated renders it particularly useful.

Wood stained black, to produce so-called ebony, may be wax polished. The result is certainly a closer approximation to the appearance of real ebony than when the work is French polished in the usual way. By polishing fretwork articles with wax they may easily be made to look better than many of them do when unskillfully French polished.

Though it has been said that any wood may be wax polished, there can be no question that this process answers best on the more coarsely-grained woods, such as oak and ash; for pine and other light woods of close texture it is not so well suited, unless they have been previously stained.

The ingredients for wax polish are, in the simplest mixture, beeswax and turpentine. Resin and Venice turpentine are occasionally added. Resin is added with the intention of hardening the surface; but provided the wax be of good quality, these additions are quite unnecessary, if not injurious, and a good result should be got from wax and turps.

Wax and turpentine alone are all the materials necessary to make a good wax polish, and when anything else enters into the composition the mixture is one of a fancy character. It is not proposed to discuss the qualities of beeswax offered for sale, and the polisher must decide what kind he gets. Some advocate the use of fine white wax, and possibly a better finish may sometimes be got with it than with the ordinary yellow wax, which, however, is the kind generally used; the only occasions when it might not be so good as the white are when extreme purity of tone is required for a light wood. Wood perfectly white is, however, seldom wax polished.

The way in which wax polish is prepared depends a good deal on the proportions of the materials. For a liquid polish, shred the wax finely, and pour the turpentine over it, leaving the two till they are incorporated. Cold turpentine will dissolve wax slowly, but a more expeditious method is to melt the wax by heat, and before it has time to solidify pour the turpentine into it. Caution is necessary when melting wax, and on no account should the turpentine be poured into the wax while it is still on the fire. With ordinary care there is no danger, and the possibility of a mishap is suggested merely for the benefit of those who might otherwise overlook the inflammable character of turpentine vapor. Should the mixture be either too thick or too thin, there will be no trouble in altering its consistency afterwards.

To thin a mass which is too stiff, a very moderate warming, by placing the bottle in hot water, will reduce it to a more liquid form, as the turpentine already in it facilitates the change, and more turpentine is added. To stiffen the mixture, wax should be melted separately, and the original mixture added to it. The heat of the freshly melted wax will probably be sufficient to cause all the materials to mix. In any case, the wax should be thoroughly melted before the turpentine is added, as a lumpy mixture is neither pleasant to work with nor conducive to good finish. The natural tendency of a wax polishing mixture, exposed to the air, is to stiffen, on account of the evaporation of the turpentine. A considerable time must elapse before there is an appreciable alteration, and the fact that a change does go on, slowly, is mentioned to remind polishers that if they have a considerable quantity of the mixture standing over, they must not expect it to retain its original consistency unless kept in a closed vessel, such as a tightly-corked bottle.

A hint for those who think that the more ingredients a mixture contains the better it must be, and who are not satisfied unless there is a certain amount of **resin** in their wax polishing paste : Always melt the **resin** first, and add the wax gradually, and constantly stir. Whether resin be used or not, the mixture should be allowed to get quite cold before it is applied to the work.

Although the consistency of wax polish varies considerably, the comparative merits of different degrees of stiffness or fluidity must be considered, so that an intelligent conception of the polisher's aim may be arrived at. Suppose a piece of beeswax, without any admixture of turpentine, is rubbed on a piece of smooth, flat wood. Some of the wax adheres to the surface, which, when friction is applied, becomes glossy or polished. The labor, however, is considerable, and though dry wax may do on a flat surface, when mouldings or carvings are to be treated, the difficulties in the way of satisfactory application are considerable. The remedy is to soften the wax so that it may be got into all parts of the work. Melted wax might do, but in putting it on to the wood it becomes cold, and consequently reverts to its original stiffness. We have then to get the wax to a fair working consistency by means of some suitable solvent, which turpentine has proved to be. It is cleanly, inexpensive, and evaporates sufficiently quickly, besides mixing well with the wax. Some polishers prefer what others might think an excess of turpentine. When a stiff paste is used, the wax is apt to be deposited in excessive quantity, necessitating a considerable amount of rubbing, in places, to remove it. A fluid polish spreads the wax much more evenly, but no gloss can be obtained till the turpentine has disappeared, either evaporated or been absorbed by the wood. When the polish has been laid evenly over the work, this does not take long; so a thin mixture may be considered preferable to a very stiff one. A paste of about the consistency of butter in hot weather, might be regarded as a medium. Those who use a wax polish which could be poured would consider this stiff, while others who add very little turpentine, or who believe in resin, would consider it thin. A thick mixture or a thin one may be used, the result depending more on the manipulation of the material than on the material itself; and this manipulation may next be considered.

In the application of wax polish there is almost as great a variety in practice as in proportion of ingredients. The great thing is to have the wax—the turpentine is merely the vehicle for conveying this—evenly and thinly distributed, and so long as this is done it is of small consequence how it is managed. To spread the wax with, some use a piece of rag, while others prefer a stiff brush specially made for the purpose, and both get equally good results. After the wax has been spread the polish is obtained by friction, and the more you rub the brighter the polish will be. The brush or cloth used to rub the wax into the wood should not be employed to give the finishing touches. In this final friction it is essential that the cloth or brush used be perfectly dry, as if it is at all damp no polished surface can be produced. The final polish is best done with a perfectly clean rubber, and three sets of cloths or rubbers may be used. With the first the mixture is to be rubbed on the wood, with the second it is to be rubbed off till a fair amount of polish is got, while with the third the rubbing should be continued till the surface is as bright as it can be got.

The directions which have been given should enable

any one to wax-polish wood successfully. Hard dry rubbing, with energetic application, is at least as important as the wax and turpentine; for, though more simple than the French polishing process, it is more laborious.

Oil Finishing and Dry Shining.—The following description of the methods of oil polishing and dry shining is taken largely from an English work on the subject, and may be depended upon as being fairly correct and up to date:

The simple process of oil polishing must now receive attention; and there is still something to admire in a comparatively dull oiled surface. The process simply consists of rubbing in linseed oil and polishing with a soft rag. The oiling and polishing must be continued at intervals till the requisite shine is obtained. To get the best results takes time and friction. Oil polishing is not difficult, but it is decidedly fatiguing and tedious. The more the surface is rubbed the better, and the process may be extended over some weeks. Patience and energetic application are still more essential than with wax polishing, for to get even the semblance of a polish or gloss within a week or two with the aid of oil must not be expected. How long does it take to finish a thing properly with oil? It may be said the work is never finished. An oiled surface will always bear more rubbing than it has had, and will not be deteriorated by friction; still from one to two months should suffice to get a good polish, which will be durable according to the amount of labor bestowed upon it during that time. This is more time than can be devoted to the finishing touches of a piece of furniture generally nowadays, so it may almost be considered that oil polishing is an obsolete process.

Still, it does not follow that because the process is too long to be remunerative in ordinary work it should not be worthy of attention, especially as it has merits which recommend it where speed is not a primary consideration. One great advantage of it is that it is much more durable than either French or wax polish; it does not blister by heat like the former, nor spoil with water to such an extent as the latter, with which in general appearance it may be compared. It is because it does not blister by heat that it is especially useful. An ordinary French polished dining-table top shows the damage caused by hot dishes laid on it, unless great care has been taken. On an oil polished dining-table top the same hot dishes might be placed almost with impunity; and it is chiefly dining-table tops that have prevented oil polishing becoming quite extinct. Though the whole of a table, or anything else, may be polished with oil, it is usual, even when the top is oiled, to polish the legs and frame otherwise.

Linseed oil is the only material used in pure oil finish, but other ingredients have been used, till it is difficult to recognize the distinction between oil polishing and French polishing. The two processes may overlap to an almost indefinite extent, but with these we have, at present at any rate, nothing to do, and to discuss them might only tend to confuse the novice. Authorities differ on the state in which the linseed oil should be used, some recommending boiled, others raw, and others various proportions of the two. For ordinary work boiled linseed oil is perhaps the better, but this is not intended to imply that those who prefer raw oil are wrong; therefore any oil polisher who has an inclination for some fancy mixture of boiled and raw oils can use it.

The treatment is very much the same as in wax polishing. It consists in rubbing the oil well into the wood, not saturating or flooding, but scrubbing it, and then rubbing long and hard. The process may be repeated almost indefinitely, daily or at longer intervals, till a polish which is deemed sufficient appears. For example, take a table top, rub some oil well into it, and then polish with a rubber formed by wrapping some baize, felt, or similar material round a brick or other suitable block, the purpose of which is, by its weight, to some extent to relieve the polisher from using his muscles in applying pressure. The rubbing should be continued till the surface of the wood is dry. The only perceptible difference in the top will be the darkened appearance caused by the oil, as little or no gloss will appear at first. By repeating the operation, however, a polish will come up gradually, and a surface which in the opinion of many is superior to that of French polish will be the ultimate result. Should the polish sweat, some methylated spirit may be rubbed in. This will dry the surface without spoiling the polish.

Oil polishing is hardly suitable for anything but plain work, on account of the labor required; but any piece of work can be so polished if the necessary time and labor be given to it. Even when it is not deemed practicable to bring up a polish with oil, a very pleasing finish may be given to a piece of work by merely rubbing it with oil. The color is enriched to an extent which perhaps would hardly be credited by those who have not had frequent opportunities of seeing wood in the white and again after being oiled. In choice mahogany especially the improvement is very marked. Light oak is also greatly improved in tone.

Fretworkers who are not proficient in French polishing would be more satisfied with the appearance of anything they make if they simply oiled it instead of coating it with shellac, which has to serve for French polish.

Dry shining will be found a simple process after the ordinary methods of French polishing have been mastered. Finishing work by dry shining is the crudest and simplest way in which a gloss can be got on the surface of wood by means of a thin varnish of shellac and methylated spirit. It must not be mistaken for varnishing, as this process is ordinarily understood, for it is distinctly a process of French polishing. Even those who have managed to do bodying-up and spiriting-off, or even glazing, will find the operation of dry shining simple in comparison. It is the nearest approach to varnishing by means of a rubber, instead of a brush, that polishers practice. The wood is varnished with ordinary French polish, applied by means of the polisher's special appliance—the rubber.

Dry shining, unlike glazing, is not in any degree a substitute for the difficult process of spiriting-off, and those who think to get a high degree of finish on their work by means of dry shining may give up the illusion. When a really good finish is wanted, French polishing, as it is ordinarily understood, should be chosen, for there is no efficient substitute by which a like result can be got.

Dry shining can be used in any position where a high degree of finish is not necessary or customary. It is useful for finishing inside work—such as the insides of boxes, drawers, cabinets, and interior parts generally—and is often seen on the fronts of drawers and trays enclosed in a wardrobe. The chief advantages in connection with it are that it can be done expeditiously, and therefore cheaply; that it sufficiently closes the grain of the wood to prevent dust getting in and clogging it; and that it gives a certain degree of finish which wood, left in the white or altogether unpolished does not possess.

The wood is bodied-in without any preparatory filling, but otherwise precisely in the manner as already directed. It is not customary to take such precautions to get up a good body as there recommended. A better description of the process is to say that the wood is wiped over with the polish rubber; not much trouble is usually taken to do more than get the preliminary body worked on. There is no reason why the first body should not be allowed to sink, and the article then be rebodied if necessary. Much bodying-in would make the work almost as hard as that involved in ordinary French polishing, so that ordinarily the bodying-in dry shining is done more quickly.

When the bodying-in has been done to the satisfaction of the polisher, the rubber is charged with French polish, rather more fully than was recommended for bodying. Instead of being rubbed all over the wood in any direction, it is wiped over in the direction of the grain from end to end of the piece, very much in the manner mentioned in connection with glazing. The rubber may be moved backwards and forwards till dry, but a better way under ordinary circumstances is to let the polish deposited by each rub dry before going over the same place again. When using the rubber in finishing, it should have no oil; and if the former of these two methods is adopted it will be difficult to prevent the polish dragging, so the easier course should be adopted.

Repolishing and Reviving.—Having once mastered the tundamental principle of polishing, it is a comparatively easy task to give to a plain piece of wood a level and lustrous surface; and, by the use of stains that can be bought ready prepared, a fair imitation of any given wood can be obtained with but little labor. But the polisher who wishes to hold his own against all comers, must be able to do more than merely to stain and polish a plain piece of new wood.

When dealing with old work that requires repolishing, all dirt, grease, and furniture paste must be removed by careful washing with soda and warm water and powdered pumice-stone or bath-brick. It can then be French polished, or a fresher and more satisfactory appearance may be given by applying one or two coats of brown hard spirit varnish—such as can be bought at an oil and color merchant's—carefully with a camel-hair brush.

When varnished work has to be dealt with, first clean off all the varnish and then repolish in the way described in previous chapters, except that filling will probably be dispensed with. The varnish can generally be more easily removed by scraping than by papering. With care the varnish can be washed off with soda or potash and water, but on account of the liability to injure the wood it is scarcely advisable to adopt this method.

For removing polish from flat surfaces, the steel scrapers as used by cabinetmakers are the best tools to use. In turned and other work which has an uneven surface the old coating can nearly all be got off by application of strong hot soda water, to which may be added some oxalic acid in difficult cases. When a large quantity of work has to be treated, use the following mixture: ½ lb. potash, ½ lb. soft soap, ½ lb. rock ammonia, 1 lb. washing soda, 3 ounces of nitric acid, 1 gallon of water. Apply with a fiber or scrubbing brush, taking care of the hands. Swill off with plenty of clean water. When the work is dry, oil and fill in; then repolish.

Spirit varnish can be removed by washing with methylated spirit, which redissolves the lac. This is both a tedious and somewhat expensive method, which need be resorted to only for delicate mouldings and other work which cannot well be cleaned by scraping or by scouring with some liquid which, though it would remove the varnish, might stain and so spoil the wood. Alcohol, being neutral, may be used on any wood, as it will not affect the color.

When dealing with cabinets or other built-up work, the process of repolishing will be simplified somewhat by taking apart as much as convenient. It is a good plan to unhinge all doors, to remove all carvings that may be screwed on from the back, and to remove all knobs, brass fittings, etc.—not forgetting to put some tallying mark on each piece which might be liable to misplacement. Thus the doors can be better handled on the bench, the corners of panels can be worked up better, and the carvings can be varnished better. When the carvings are planted on, as is often done, a much cleaner job is made if these are first removed; for it is a difficult task to polish the open carvings equal to the flat surface.

Sometimes polished work is disfigured by fine little lines which are caused by cracks, resulting from sweating. These lines become visible through the dust settling on the exuding oil. This disfigurement can be averted almost entirely by occasionally carefully wiping with a soft, damp cloth. Sweating is not entirely preventible, but when the oil has ceased to exude, which may not be for some months, the work may be repolished with advantage.

The perfectly level, brilliant polish found on new German pianos fills many an American French-polisher with envy. Unfortunately, this brilliant polish does not last long, and the majority of the pianos soon have a greasy, cracked appearance. Indeed, there are but few of these pianos with a polish gained by the legitimate process of French polishing. This brilliant, level polish is gained by a very liberal use of gum sandarac, and when the polishing is completed the pianos are set aside in a clean, hot room, which has the effect of causing the polish or gums to flow to one dead level. Some makers use varnish very freely, and, before passing to the hot room, level this by means of pumice powder, tripoli, putty powder, and sometimes flour.

When the requisite number of coats of varnish have been laid, the surface is leveled with fine glass-paper and linseed oil, or by the slower process of felt rubber and pumice powder. After being wiped perfectly clean, a rubber made of soft flannel, or, better still, of old silk, is used to rub carefully and lightly in a circular direction with tripoli powder and oil, till the surface is perfectly level and inclined to be bright; it is then rubbed with dry putty powder and silk, and finally brightened with flour.

The surface should be left perfectly free from any trace of the polishing powders; neglect of this accounts for the white patches sometimes seen on the German pianos. These patches are not so deep as they appear at first sight, and may often be removed with flour emery and linseed oil or turps without disturbing the polish.

To renovate the polish on these pianos is difficult, but when it is not very bad, a reviver made of equal parts of linseed oil, lime water, and turps is generally effective. The lime water and oil are first thoroughly mixed; then the turps is added, and the mixture is applied by means of wadding. The surface is wiped off with a rag, and finished with a clean, soft rag-swab, made fairly moist with methylated spirit. Should any trace of grease still remain, change to a clean place of the already moist rag, and sprinkle a few spots of glaze on its face, or, better still, wipe the face of the glaze rubber over the face of the clean swab.

Should this method prove ineffective it will be necessary to repolish, first removing the sweat or roughness by fine sandpaper and oil, or by washing with weak soda water and pumice powder. The polish used should be made with spirit instead of naphtha, and, to ensure its lasting qualities, it should be bodied up one day and finished the next.

To darken the birch frame of a chair, wipe it over with asphaltum dissolved in turpentine (one cent's worth in half a pint of turps). This stains without giving a painted appearance; should there be any difficulty in obtaining asphaltum, Vandyke brown may be used, mixed to a thin paste with liquid ammonia— or with a strong solution of common washing soda. This is thinned with water, till of the required tone, which will readily be found by trying its effect on any odd piece of wood. If French polish cannot be applied, the most suitable thing to use is brown hard spirit varnish.

For restoring polish that has faded from damp or

exposure to the sun, those stains which are used to stain the common woods will not be suitable. It may be convenient to remove only the upper surface of the polish, to color the faded portion so as to match its surroundings, and to repolish the whole. When the polish is not very bad, it is generally sufficient to smooth it well with a piece of worn glass-paper. When it is much scratched or faded, methylated spirit should be sprinkled upon it, and the surface well rubbed with No. 1 sandpaper, applied with a circular motion; it will then be found that only the upper surface of the polish will be removed. Before repolishing, it is advisable first to wash the article with water to which a little common washing soda has been added. This will remove any dirt, furniture paste, etc.; a little pumice powder or powdered bath-brick may be used to assist.

After the necessary cleaning off of dirt, etc., has been accomplished, any bruises must be removed, either by scraping out or by bringing up level, by means of a hot iron and moisture, or by filling up with hard stopping, or by the still better method previously given. When this has been done, and all defective parts made good, the surface must be wiped over with an oily rag; it assists the new polish to take kindly to the old. In scraping out the bruises, in cleaning off level any new piece, and in cleaning off the polish, it is probable that light patches may be made. More especially will these be made apparent if the damaged portion has been previously colored up by stains, dry colors, or dyed polish.

For coloring up or matching, it is generally suffi-cient, if the wood in hand is mahogany, to wipe over the damaged portion with red oil, which consists of ¼ lb. of alkanet root steeped in 1 pint of linseed oil, working up with red polish. Should the wood be walnut, many a little blemish and scratch in soft resinous varnish may be matched by wiping over with a solution of one cent's worth of asphaltum dissolved in ½ pint turps. Should the defect be a piece of sap or other light portion, go over the light portion several times with the polish rubber to prevent the grain from rising, and then saturate a small tuft of wadding with 3 parts of methylated spirits to 1 part of polish; on this wadding place a small quantity of Vandyke brown or brown umber, mix well, and carefully wipe over the light portions, thinning out with spirits if too dark, picking up a little more color if not dark enough, adding a little black if required.

Matching stains are used in French polishing because light and dark places often occur in the best selected woods, and in stained work, owing to the difference in the direction of the grain. To tone or harmonize the entire surface to one uniform shade, is technically called matching or coloring up, and requires a little tact and a good eye for color. On large flat surfaces colored polish may be used with advantage, but for small work it would be better to take a small tuft of wadding and wet it with 1 part polish to 3 parts spirits. With this take up a little yellow ocher and just a trace of umber or Vandyke brown. Press the wadding well on the back of a piece of worn-out glass-paper to equalize, and mix well. Try the effect on an odd corner of the work; if too dark, thin out with spirits; if not dark enough, pick up more color, or wipe over twice. Having gained the right shade, apply lightly with a straight or wavy motion as required. This would enable one to match the oak, but any wood can be matched by using suitable pigments, a red tinge being usually given by the addition of a few drops of Bismarck brown stain. Though it is possible to proceed to polish direct, yet it would be safer to set the stain by giving a coat of thin spirit varnish, and allow this to get quite dry before polishing. In matching up satin walnut, the polisher must use judgment, for the work can hardly be regarded as mechanical. The stain must depend on the color or tint of the lighter parts, and of the darker parts to which they are to be matched. Generally a little weak brown stain will do what is required. When necessary, it can be altered slightly in color by the addition of other pigments, according to the tints desired.

In matching, the wavy appearance of some woods may be given by a tremulous movement of the hand, and the mottled appearance of others by dabbing with a badger softener or clean, soft dusting brush while the color is still wet. Veins either black or red may be given by picking up a little dry black or red stain on the corner of the tuft of wadding and applying it carefully, taking some adjacent portion as a guide for pattern. For rosewood, red stain and dry black may be used in combination; for birch or oak, use yellow ocher. When the work in hand is large, and requires staining all over, and it is not possible to gain the desired result by means of dyed polish applied with the rubber, the colors should be mixed in a pot with 3 parts of spirit to 1 of polish, and applied with a camel-hair brush. The work is not so liable to get patchy with two or more coats of weak stain as with one strong one.

After laying on the stain, allow a few minutes to elapse for it to set, then smooth down with a piece of worn, fine glass-paper, and give a coat of thin brush polish or spirit varnish. This will set the colors previous to polishing, which can be proceeded with in about ten minutes. Mahogany, rosewood, and walnut, if not inlaid, are generally improved by the use of a polish tinged by the addition of a little red stain. Other colors may also be mixed with polish to be applied with the rubber. When using these dyed polishes, cease when just the right tinge is attained; another rubber may be used to finish off with clear polish.

In repolishing work, the foundation having been already laid, the polish is not required quite so thick as in polishing the bare wood. In the final stage, when finishing off, any trace of greasiness may be effectually removed by well rubbing with a swab of the clean, soft rag, fairly damp (not wet) with spirits, on face of which has been sprinkled a few drops of glaze.

Colors in a dry state known as pigments, such as Venetian red, yellow ocher, vegetable black or lampblack, umbers, Vandyke brown, chromes, orange and lemon, greens, blues, flake white, etc., are useful. By the aid of these, the polisher is enabled to match woods and restore faded polish, far more expeditiously than can be done by staining or using dyed polish or varnish. Work that might puzzle the inexperienced for hours can be done in a few minutes by a knowledge of the use of dry colors. They are used in some stains by mixing with ammonia, glue size, pearlash, soda, and they are used to color the "filling-in" of whiting and turps to make it match the various woods. Venetian red is used for mahogany, umber for walnut, black for ebony, and sometimes to give an appearance of age to oak by making the grain appear dirty.

Sometimes the polisher has a job passed to him that, properly speaking, ought to be done by the painter. The quick-drying nature of the solutions of shellac, with which the polisher is the better acquainted, obtains for him the preference. For example, shields, etc., for the decorative purposes that may require five different colors and a coat of varnish, can be colored ready for fixing within twenty-four hours. To do this, lime blue, chrome yellow, vegetable black, flake white, and vermilionette or any other colors should be mixed with ordinary French polish to the consistency of thin paint, thinning out when necessary with alcohol spirits. Three coats of color can be laid on, stencil patterns cut and painted, borders and edges lined, and the whole finished with a coat of white hard varnish within twelve hours. To prevent the white getting a yellowish tinge, it is well to mix it with transparent polish made from white shellac.

To make imitation marble which wears well, give several coats of flake white mixed in polish; then put in the veins of blue or black with feathers, afterwards giving a coat of white thinned out with spirits. This has the effect of making the veining appear beneath the surface. When dry it is finished by giving a coat of good quality copal varnish.

THE FINISHING OF VARIOUS WOODS

It will be in order now to describe how to finish some of the more popular woods, in detail, though enough has been said in a general way to enable any workman to finish work in any wood made use of for building or decorative purposes, but a few lines devoted to some of our special woods may perhaps be of timely service to some owner of this volume. Oak, being the most popular and at the same time the most durable and the most used of our woods, is deserving of the first place.

Finishing Oak

We have in this country over forty kinds of oak nearly every one of which may be used for some special purpose; those most used for building finish and for furniture, however, are only few, among which are the white oak (*Quercus rulia*), rock oak (*Quercus primas murticola*) and black oak (*Quercus tindoris*). All of the above-named oaks are capable of being handsomely finished, the white and black oaks being the best and the red being next.

What is known as quarter oak is made by first sawing the log from end to end through the middle. Then each half is sawed from end to end through the middle, thus leaving four quarters. Each quarter has only three sides, one side the bulge part of the log, and the other two sides flat and coming to an edge. The boards are sawed off the sharp edge, and each sawing, therefore, throws off a board wider than the one before it. Sawing the quarters of the log in this manner, lumber possesses that beautiful cross-grained figure so much in fashion now that it has become somewhat of a craze. This cross-grained material finds favor in the finest furniture and interior work. The wood is susceptible of the very finest polish, and the cross-grain produces an effect made by both nature and the saw, that is quite superior to the art of the most skillful grainer. To effect a good imitation of antique oak, lampblack or Vandyke brown in oil is applied to the surface of the wood, darkening its natural hue; but this is not by any means best or the

only way, but answers very well where other processes
or methods are not available. With regard to giving
oak an antique appearance, many opinions exist.
Formerly—and probably the first—imitations of antique
oak were produced by exposing the bare dressed surface
to the steam of boiling ammonia. This process, however, it was impossible to apply to casings of house
interiors, thus leading to the application of the aqua
ammonia with a brush, but it is by repeated application
that the desired results are produced. It stains the
wood gray. There are finishers who claim this to be
the only process by which to imitate Nature, but
another, much faster and one which gives the same
results, is to use strong vinegar with iron filings or
shavings added; by a little experimenting this can be
made to suffice with one coat, depending upon the
amount of iron added.

A very clever imitation of the general antique can
be obtained by staining the filler with Vandyke brown
and charcoal, equal parts, using about 1 part of the
colored to 4 parts of the light. Then there is another
antique which imitates certain oak from the sixteenth
century; the peculiarity of it consists in dark cloud
streaks permeating the wood in every direction, some
of them crossing the panels in a V-shape, others
straight near top and bottom. To give them an odd
appearance, a pair of panels can be clouded by
streaking one three or four times and the other once or
twice. These stripes vary in width from three to five
inches. The wider streaks look well across the top of
a table three or four times, or even partly across, while
the narrow ones would do well around the legs and
across the styles and rails of paneled work. In
putting these on, they should be dark in the center and

blend out at the edges. This is done with an automatic paint burner, allowing the flames to scorch the wood nearly black; but care must be taken not to char it. After the work is all streaked, the wood is filled with filler stained with burnt umber mixed as for black walnut

Care must be taken to have the filler stained to the proper tint, and in applying correctly. By following the rules given under the head "Filling" the operator should have no trouble in making good work, and we here repeat the instruction.

After enough surface has been covered with the filler, so that what has been first applied begins to flatten, the process of wiping should immediately begin, using for that purpose either a rag or a handful of waste or excelsior. If the oak is very open-grained, waste is preferable. With a piece of this that has previously been used and is pretty well supplied with filler, rub crosswise of the grain, rather rubbing it into the grain than wiping it off. After the whole surface has been gone over in this way, take a clean piece of waste or rag (never use excelsior for wiping clean) and wipe the surface perfectly clean and free from filler, using a wooden pick, the point of which has been covered with a rag or waste, to clean out the corners, beads, etc. It is well to give these picks some attention, as a person once accustomed to certain tools can accomplish more and better work than with tools that feel strange in his hands; therefore, each finisher should furnish his own pick. As to their construction, those are best made from second-growth hickory, which can be procured from any carriage repair shop, such as old spokes, broken felloes, etc. They are made eight inches in length, half inch oval at one end and tapering down to the point at the other. Sharpen the oval end like a coal chisel, then smooth with sandpaper, which should also be used to sharpen the tool when the same becomes worn dull.

This picking out of the filler from beads, etc., can be accelerated by the use of a picking brush manufactured especially for that purpose, but it is not advisable to use this on very coarse-grained oak, as it scrubs the filler out of the pores.

Oak may be fumigated by liquid ammonia, strength 880°, which may be bought at any wholesale chemist's at $1.50 a gallon. The wood should be placed in a *dark* and *airtight* room (in a big packing case, if you like!), and half a pint or so of ammonia poured into a soup plate, and placed upon the *ground* in the center of the compartment. This done, shut the entrance, and secure any cracks, if any, by pasted slips of paper. Remember that the ammonia does not touch the oak, but the gas that comes from it acts in a wondrous manner upon the tannic acid in that wood, and browns it so deeply that a shaving or two may actually be taken off without removing the color. The depth of shade will entirely depend upon the quantity of ammonia used and the time the wood is exposed. Try an odd bit first experimentally, and then use your own judgment.

Short pieces of stuff may be so treated by using an airtight box. The box ready, a flat dish or plate of strong ammonia should be placed in the bottom, so that the fumes will *rise* and surround the object. All that is now necessary is to place the article in the box, nailing up as close as possible, and await results Ten hours' exposure, using strong ammonia, should give a good color; if not dark enough let it remain longer, bearing in mind, however, that the wood will present no noticeable change until oiled or brought in contact with a wet substance such as shellac. It is well, therefore, to note the progress by touching the wood with the wet finger, when it will show at once the stage it has reached.

There could be no better method devised to stain oak than this, when practicable, and in adopting it we simply anticipate nature, which, in time, through the action of the ammonia of the atmosphere, would present the same result. Mahogany may also be treated similarly with success.

Here is another method of making antique oak, and it might be added that white, and black ash, and chestnut, similarly treated, will give a fair imitation of antique oak. The job should be made of hardwood, with as full an open grain as possible to secure a fine effect. Sandpaper this and clean off. Then prepare a priming made of 1 part japan, 1 part raw linseed oil and 1 part rubbing varnish. Drop into ½ gal. of the liquid 1 lb. of commercial corn starch, such as is used for culinary purposes. Next take some good, dry, burnt Turkish umber, and add about ¼ lb. of this to the starch. Apply to the job a good flowing coat of this priming. Let stand until it is set and has soaked well into the grain, and then take a broad putty knife and stick it into the grain, working the knife crosswise of the grain. Again let stand a little while, and then wipe with rags; especially clean out all the corners, and get the job into as good condition as possible as regards having the grain well filled.

Upon the completion of the operation above described it will be found that the open grain has absorbed the starch and umber, and that these portions now show the dark shade suggestive of age, while all the rest of the surface is also slightly darkened.

When again perfectly dry, give one coat of rubbing varnish, prepared by adding to it ½ lb. of starch to each gallon of varnish. This coat should be flowed on freely as a medium coat of rubbing varnish, but be careful not to have runs of sags. This ought to completely fill the wood, after which proceed to varnish, rub and finish the job in the usual manner. To produce a natural oak finish, follow precisely the same course as above described, with the single exception of omitting the umber. This will leave the wood in its natural color.

Some of the most attractive work in this line, however, is effected by simply spreading on the surface of the material a concentrated solution of permanganate of potash, this being allowed to act until the desired shade is obtained. Five minutes suffice ordinarily to give a good color, a few trials indicating the proper proportions. The substance named is decomposed by the vegetable fiber, with the precipitation of brown peroxide of manganese, which the influence of the potash, at the same time set free, fixes in a durable manner on the fibers. When the action is terminated, the wood is carefully washed with water, dried, then oiled and polished in the usual manner. The effect produced by this process in several woods is really remarkable. On the cherry especially it develops a beautiful red color which well resists the action of air and light, and on the other woods it has a very pleasing and natural effect.

Along with the foregoing may be added the following stains for oak: add to a quart of water 2 ounces each of potash and pearlash. This is a very good stain, but it should be used carefully, as it blisters the hands and softens brushes. The stain may be made lighter by adding more water.

To Darken Oak.—To darken the color of oak, any of the following may be used:

Liquid ammonia laid on evenly with a rag or brush will deepen the color immediately, and it will not fade, this being an artificial production of result produced naturally by age.

Bichromate of potash, dissolved in cold water, and applied with a brush, will produce a similar result.

A decoction of green walnut-shell will bring new oak to any shade or nearly black.

Another.—Two quarts of boiled oil; ½ lb. of ground umber, mixed in oil by colorman; 1 pint of liquid driers, stirred in; 1 pint of turpentine; mix. After cleaning and planing your boards, lay this on with the grain of the wood. If required lighter, add naphtha till the required shade is attained; it darkens with age. Give it twelve hours to dry; then varnish with wood varnish, or use only beeswax and turpentine. The result is good in time, but slower than varnish.

Oak can be fumigated, by making a tent of some cheap oil cloth, which may be rigged up over a rough wooden frame. Of course, the tent must be made pretty tight. Don't let the frame touch the work, and when complete, cut a small piece that you can lift up and use as a peep hole. Then get a saucer full of liquid ammonia and place inside the tent—anywhere on the floor will do. Close the tent and await results. The more ammonia used the darker it gets, so you must use your peep hole and suit yourself. If you wish to polish it, give it a coat of beeswax and turpentine, let dry, and then brush or rub it well. For a dull polish, give it a coat of raw linseed oil, dry well, and then brush up. By "fumigation" you always get a good even tone; but, failing this, you can stain the work, and by simple means. Buy a little permanganate of potash at the chemist's, dissolve in water, and put on with a brush; when dry, give another coat if not dark enough. Another method is to get ½ lb. of black japan and dissolve in ½ pint of turpentine, and apply as before, or you can buy the stain ready-made; and the walnut *water* stain—not varnish stain—is a good one to make a dark brown oak color. If you find the stain brings up the grain—i.e., makes the wood rough—rub it down with fine sandpaper and stain again, but rub off fairly dry. All these are simple methods, which you could easily acquire.

Styles of Oak Finish.—As oak in its many and varied finishes is so very fashionable just now, an explanation of the effects of the several stains may be of interest, as so many of them are so closely allied as to confuse the uninitiated. Bog oak is a thin stain of medium color, giving quarter-sawed oak a slight tinge of green. It is about the same density as weathered oak, but of a green tone instead of brown, like weathered oak. Weathered oak is of a brown tone in close imitation of the rich old hue taken on by oak through time or from exposure to the weather. Antwerp oak is also brown, but of a deeper shade, producing an attractive antique effect. Black Flemish is a much-admired finish, especially when it is desired to produce an effect of great weight. It gives a piece of furniture a substantial appearance. Its black tone combines admirably with red wall covering and hangings. Brown Flemish is not unlike Antwerp, but of a much stronger brown tone. This is one of the most popular stains of the hour. It is quite permanent and produces a very artistic effect. The so-called gun-metal finish for oak is not unlike black Flemish, but gives a tinge of blue instead of deep black. Malachite, although light green, is not too intrusive. This is affected by many people of good taste and is quite popular, especially for staining ash. Tyrolean oak is as dark as black Flemish or bog oak, and is of a green tinge, instead of the blue of the gun-metal finish.

Golden Oak is very fashionable, being a brownish color with a sort of golden glow tint; it may be obtained as follows: Golden oak finish is not produced by the filler alone; in fact, the filler has very little to do with the result. The wood must be stained before it is filled, and, of course, the filler must be so colored or stained as not to mar or dull the effect. A mixture of gold size japan and genuine asphaltum varnish in about equal parts, thinned with turpentine, makes a good stain that will not raise the grain of the wood, dries quickly and hard, and, if wiped out properly, gives under varnish a rich effect, termed "golden," for want of another appropriate name.

To make a filler, mix one-third each of raw linseed oil, japan gold size and turpentine, and put into this mixture enough finely powdered silica or silex to make a stiff paste, and color this with burnt umber in oil, Vandyke brown in oil and a trifle of drop black to suit, being mindful that in golden oak only the high lights are yellowish brown, while the filled grain is decidedly dark. The mixture should be run through a handmill. The best plan for you is to buy your golden oak paste filler, or at least buy the light paste filler and color it to suit your taste; for you cannot buy the raw material as cheap as the manufacturer, and making it in a small way will cost you more in the long run.

Another method, if the work is new, is as follows: Fuming is only possible when the articles are new and free from varnish, polish, glue, or marks of handling. The process consists of enclosing the articles—from which the glass and all brass fittings are removed—in an airtight room or box, on the floor or bottom of which are placed a number of shallow dishes containing strong liquid ammonia. The depth of color depends on the length of exposure, which may vary from twelve to thirty-six hours. Where this process is not practicable, the next best method is staining. The stain must be weak, the exact color required being afterwards obtained by the use of a polish made from orange shellac and a trace of color in the varnish. A suitable stain may be made by dissolving ½ oz. of bichromate of potash in 1½ pt. of water. To prevent the work coming up rough on the application of the water stain, the work should be first wiped over with raw linseed oil. The stain must be liberally applied, and rubbed well in with a rag, finishing off always in the direction of the grain. Before starting on the work, experiment on odd pieces of similar wood.

Flemish Oak.—To make a stain for Flemish oak, ½ lb. of bichromate of potash, dissolved in 1 gal. of water. Coat woodwork. When dry, sandpaper down smooth; then coat with best drop black, ground in japan, thinned with turpentine. Let stand five minutes and wipe off clean, then coat with pure grain shellac and sandpaper with No. 0 sandpaper; then coat with beeswax, 1 lb. to a gallon of turpentine, ¼ lb. of drop black mixed in the wax, then wipe off clean with cheese cloth.

Weathered Oak.—Give woodwork one coat of strong ammonia. When dry, sandpaper down smooth and stain it from the following colors: lampblack, ochei and 2 lbs. of silica to a gallon of stain. Wipe off with cheese cloth, then give one coat of pure grain shellac, then sandpaper and give one coat of wax and wipe off clean. If you should desire a brownish shade, put 1 oz. of bichromate of potash and ammonia, or if a greenish shade, put some green and stain.

Verde, or Green Finish.—One ounce of nigrocene dissolved in ½ gal. of water. Give woodwork one coat; when dry, sandpaper, care to be taken not to rub off edges; then fill with a bright green filler, with some white lead in the filler. When thoroughly dry, give one coat of pure grain shellac and then wax, or it could be finished with three coats of varnish and rubbed. This finish leaves the pores of bright green color, while the rest of the wood is almost black.

Black Oak.—One ounce of nigrocene to ½ gal. of water. Give woodwork one coat, then fill with a black filler, then one coat of shellac and three coats of varnish rubbed with pumice-stone and water, then oil and wipe off clean.

Austrian Oak.—Fill with a light antique filler, colored with raw umber. Give two very thin coats of shellac, colored with nigrocene and yellow to the desired shade, then sandpaper down and wax and wipe off clean.

Red Oak is a difficult wood to stain or fumigate, but it may be done as follows: Make a stain by mixing ground dry Dutch pink (this color is yellow) and a little dry drop black, with beer, and apply with a hog's hair brush. Try the stain on a piece of red oak, and get the exact shade if possible, taking care that you do not stain quite as dark as the fumigated parts. When dry, oil with linseed oil, and make a weak stain, using the same colors, mixed with alcohol, with enough button polish to bind the color. Lay on carefully with a camel-hair brush. It is better to give two coats of weak stain, as the result will be a more even color. When dry, wax polished this will make a good job.

Oak Staining Generally.—There is no wood which may be treated in so many different ways as oak. It may be left in its natural state, or it may be oiled, or wax polished, or French polished in its natural color. It can be stained and waxed, stained and French polished in a variety of colors or tints, with the grain opened or filled, and it can also be fumigated. For bed-room furniture, if the wood is carefully selected, a very pleasing effect is obtained by waxing or French polishing it in its natural color, or slightly stained and polished with the grain open or filled, according to taste. Dining-room or library finish is generally stained a medium color or fumigated. All furniture is frequently stained very dark, and polished with the grain open.

If it is desired that the work should be finished in its natural color, fill in with Russian tallow and plaster of Paris, and polish with white polish. If it is required to be slightly tinted, stain the filler with yellow ocher and polish with button polish. For staining, the best stains to use are the powdered water stains, and some very effective tints may be obtained by carefully mixing green and brown stains together; apply the stain with a hog's hair brush, and if the grain should rise quickly, rub down with 1½ glass-paper before laying off the stain. When the stain is dry, oil with linseed oil, then give a coat of polish to fix the stain. It may then be polished with the grain open, and finished with a wet rubber, using no spirits. If a level surface is required, it must be carefully filled in, and not unduly hurried in the polishing. The latter applies to oak generally, as the wood is coarse, and consequently sinks a great deal.

A good dark oak stain may be made as follows: Dissolve 1 oz. of bichromate of potash in ½ pint of water, and 1 oz. of potash in ½ pint of water. When each are separately dissolved, mix together, and add dry powdered Vandᵥke brown. If a very dark color is required, add also a little powdered drop black; apply with a hog's hair brush, and polish as before stated. As sometimes the American potash varies in strength, the hair of the brush will curl up if it is too strong. If this occurs, add a little more water. Oak carvings give a much nicer appearance always if the grain is left open, even when the other parts are filled up. If they are stained, oil and afterwards give a coat of polish to fix the stain. When this is dry, brush well with a stiff-haired brush and rub with a dry cloth. It is sometimes advisable to oil oak before it is stained, as it often prevents the grain rising. The only disadvantage is that a little extra labor is required to make the stain bite.

Fumigated Oak.—The best kinds of oak for fumigating are the English wainscot, or Baltic. The red American oak does not fumigate well. The advantages of fumigating are that a more natural color is obtained than by staining. The wood is not made rough by the operation, and there is also a great saving in labor. The best method of fumigating is to construct an airtight chamber, lined with wood, and the joints of the wood made airtight by pasting paper over them. If the chamber is of the following dimensions it will be found large enough for most purposes: length, 9 ft.; width, 4 ft.; height, 6 ft. Portable inner frames may be made with shutters, so that the size of the chamber may be made smaller if necessary. The door should have glass panels; this will permit of the work being watched, and when the wood has become dark enough, the door should be opened. Articles to be fumigated should have all brass work removed; then place in the chamber in such a position as to allow a free passage for the fumes to get at all parts of the wood. Then place half a dozen saucers (flower-pot saucers will answer for this purpose) on the floor at equal distances, and pour into each saucer ¼ lb. of spirits of ammonia, strength of the ammonia to be 880°, then paper over the joints round the door. The wood will darken much quicker in hot weather. If a very dark color is required, it may be necessary to recharge the chamber after twenty-four hours, but a good color is generally obtained in about five hours. It should be noted that the work always becomes lighter after it is taken out of the fumigating chamber. Consequently, the work must be proceeded with directly it is taken out. If any parts are too dark, do not oil them; all other parts should be at once oiled, and given a coat of polish. When dry, paper well with No. 1 glass-paper, and wax polish or French polish with the grain open, as in the case of stained dark oak. Small articles may be fumigated by making a box airtight, and placing a piece of felt upon a level floor; stand the articles to be fumigated on the felt, and fill one saucer with ¼ lb. spirits of ammonia. Then cover the whole with the box, and place a weight on the top of the box; this will prevent the fumes from escaping. It may be raised occasionally to see how the work is proceeding. This plan will answer better than if there is a lid to the box, as the fumes will not escape so much in this way.

Great care must be taken when using the ammonia, and the operator must particularly avoid inhaling the fumes. A good rule when charging a large chamber for fumigating is to have some one at hand in case of any accident, such as the breakage of the vessel containing the ammonia. Before the work is taken out of the chamber the fumes must be allowed to pass off by opening the door for a few minutes before entering. If this rule is carried out, no possible harm can happen.

Pollard oak is best treated by first oiling it, and then applying a coat of button polish. When the polish is dry, it must be rubbed down with No. 1 glass-paper and waxed or French polished. As sometimes pollard oak has very large cracks on the surface, these must be well stopped with wax stopping, which must be stained to match the wood. Before proceeding to wax or French polish, if the joints do not match in color, a little polish stain may be applied with a camel-hair pencil before it is polished.

For Removing Polish and Restaining Oak.—Saturate table with alcohol, keep it wet with it, and whilst wet scrape off polish which will have become softened. If legs are turned, or on shaped edges, etc., where scraper cannot be used, coarse sandpaper (Middle 2) will remove polish. Use plenty of spirit and sandpaper all over, and take care all polish is removed. It can be done in same way with potash—a slower process, and the potash will also burn anything it touches, but will stain the wood at the same time. When all polish is removed, the table can be stained dark with *walnut water* stain. When wood is well stained in pores, wipe off with cloth and let dry; if not dark enough, give another coat. Another stain is made with turps and black japan, well mixed. A little Prussian blue powder will make either stain a greeny brown. For polishing, rub in raw linseed oil; let dry, and rub again; or for wax polish, melt beeswax on slow heat in a galley pot; add turpentine, about a third part; let cool. Wax should be soft as paste; if sticky, add turps; rub well into wood. Let dry and rub again for polish. Stain must be quite dry before oil or wax is put on.

To Finish Cherry

Cherry (*Prunus cerasus*).—This is a fine-grained wood, tough and light; is capable of taking the very finest finish. Is harder than baywood, and is a nearer approach in color, grain and texture, to mahogany than any other native wood.

One of the best methods for making cherry look like mahogany is to have the wood rubbed with diluted nitric acid, which prepares it for the materials subsequently applied. Afterwards to a filtered mixture of 1½ oz. of dragons' blood dissolved in a pint of spirits of wine is aded one-third that quantity of carbonate of soda. The whole, constituting a very thin liquid, is brushed with a soft brush over the wood. This process is repeated with very little alteration, and in a short interval of time the wood assumes the external appearance of mahogany. If the composition has been properly made the surface will resemble an artificial mirror, and should this brilliancy ever decline it may be restored by rubbing the surface with a little cold drawn linseed oil.

When cherry is nicely filled and rubbed well down and not varnished, it has a soft glow not possessed by any other, and has none of those distortions of grain that are so unpleasant in mahogany. The timber is chosen from the wild cherry, which in New England and the North generally does not usually grow to a girth of more than 20 inches, but in some of the Western States and in the South frequently attains a diameter of 24 inches. The domestic fruit cherry gives some good specimens of small timber, but as the tree is rarely sacrificed until it is past bearing and is decayed, this source of supply is precarious. The facility with which cherry can be worked makes it a favorite with the cabinetmaker and the house-joiner; and it also possesses the quality of "staying where it is put," and that is more than can be said of many of the hardwoods.

I give below several stains for making pine and other suitable woods to have an appearance of cherry.

1. To prepare this color in water stain, boil in a gallon of water 1 lb. of Spanish annotto and 1 oz. of concentrated lye (potash). Should this not be deep enough, allow the water to evaporate by a gentle heat. The stain can also be darkened by adding gamboge previously dissolved in a weak potash solution.

2. Gamboge in oil, diluted with turpentine, and a little japan added as a siccative. This produces the same color in oil as the former in water stain, and can be deepened with dragons' blood in oil or finely ground burnt sienna in oil.

3. Mix together, by stirring, 1 qt. of spirits of

turpentine, 1 pt. of varnish, and 1 lb. of dry burnt sienna; apply with a brush and after it has been on about five minutes wipe it off with rags. This stain takes about twelve hours to dry.

4. Take 1 qt. alcohol, 2 oz. of dragons' blood; pulverize the latter along with ¼ oz. of alkanet root; mix and let stand in a warm place a couple of days. Shake frequently in the meantime. Apply with a sponge or brush. Two or three coats may be required. This makes a fine stain.

To finish cherry, the first and a very important thing to do is to give the wood a thorough sandpapering, to remove finger and other marks, and make a perfect surface to receive the stain. Next comes the dusting off of the work, which also should be carefully done, as specks of dust or dirt will cause bad work. Stain should be put on with a four-inch oval brush, one set with cement. Apply the stain freely, but do not work it too much, as this would cause it to froth, forming specks. Have the stain in a wooden, earthen or enameled vessel, as metal will often alter the color of the stain. Avoid laps when staining; do a section at a time. But should a lap be unavoidable, then take a sponge, wet with clear water, and wet that part of the work already done and adjoining that which is to be done, and then at once apply the stain. Have a bucket of water and a sponge ready at hand. Any part of the work taking too dark may be toned down by means of the wet or damp sponge, causing it to match the other and lighter work. Allow the stain to dry thoroughly, after which it is ready for sandpapering with ooo paper. Next give a coat of shellac. Finish with two coats of varnish, or with three coats for extra fine finish. Rub with pumice-stone and water, polish with rotten stone and water, and clean up with furniture polish.

Oil stains were formerly used, but aniline stains give much better color effects. Aniline stains may be bought ready prepared.

When sandpapering cherry be sure that you do not cut through, as it would show up white. Cutting through is liable to occur about mouldings, edges, etc. Use old, worn sandpaper there; for the more sunken parts redampen and rub the layers of paper from the back of old sandpaper, which will make it very pliable and soft.

Finishing Black Birch

Birch.—*Betula Nigra* or Canadian birch is superior to all other birches for constructive purposes, and when properly finished has a fine, quiet, refined look that commends itself to all lovers of domestic woods.

Black birch is a close-grained, handsome wood, and can be easily stained to resemble walnut exactly. It is just as easy to work, and is suitable for nearly, if not all, the purposes to which walnut is applied. Birch is much the same color as cherry, but the latter wood is now scarce, and consequently dear. When properly stained it is almost impossible to distinguish the difference between it and walnut, or cherry, as it is susceptible of a beautiful polish, equal to any wood now used in the manufacture of furniture and inside finishings.

To finish to represent mahogany, coat with a weak solution of bichromate of potash, then stain with rose pink, Vandyke brown and burnt sienna; then shellac, with a little Bismarck brown dissolved in the shellac. This makes a better stain and more lasting than a water stain.

There is a species of bird's-eye birch, but it is very scarce. An evidence of the weight and solidity of the wood is the fact that it will sink after being a few days on the water. It also possesses the quality of durability in a preëminent degree.

Birch is generally finished the same as cherry, and directions given under that head will apply here also.

Finishing Mahogany

Mahogany, cherry and black birch require about the same treatment, though there are some features in mahogany that differ a trifle from the other two woods. There is little or no grain markings in cherry or birch, while Spanish mahogany may be marked and penciled over its whole face; and the texture of the woods is very different to the touch.

Mahogany (*Svietened*).—The tree has a darkish-brown bark and a reddish-brown, coarsely fibered, streaky, hard wood. The tree grows to the height of 35 meters, and is pretty strong. Among the chief varieties is the common mahogany, with a very hard, very durable wood, which is never attacked by worms, and is excellent for ship-building; but its capability for taking a fine polish is its chief recommendation. *Mahagoni Haiti, Mahagoni Jamaika, Mahagoni Havanna* are the other chief kinds.

With perhaps the exception of our oaks, no wood possesses like advantages of combined soundness, large size, durability, beauty of color and richness of figure. So, when compared with other woods, mahogany costs no more to work and stands better than any other—the only point to weigh against this last great feature is the slight difference in the first cost of the wood in the rough; but if mahogany stands better and longer, and needs no attention afterwards, surely the sole advantage of less cost at first which any other wood may possess is overcome.

But another merit, equal to any thus far mentioned, is the warmth in its color and the glory in the figure of this beautiful wood. The air of elegance, artistic effect and gentle breeding it imparts to all its surroundings, its joy and life—all these cannot be measured by a few cents a square foot. Its growing splendor with age that gives increasing satisfaction may safely be contrasted with the tameness of other woods, which, though pleasing at first, deteriorate rather than improve.

When the real wood is used, but little more is necessary than to fill and varnish or polish, as it cannot be much improved upon. Sometimes, however, it may be deemed proper to darken it somewhat to take away the reddish hue that newly wrought mahogany presents, and this can best be done by darkening the filler, to suit the taste, trying the mixture first on a piece of the dressed stuff, until the desired shade is obtained. Staining the varnish or polish with dragons' blood or other suitable dyes, will also accomplish the desired end.

Staining by the fumes of ammonia will probably give the best results, as almost any tinge can be given the work, from the newness of youth to the mellowness of extreme age. This method is considered the best for imparting to mahogany the appearance of age, and for those wishing to avail themselves of an easy, clean and certain means of gaining the result, fumigating offers no serious obstacle to its accomplishment, the articles necessary being easy of acquirement, and at small expense.

To darken mahogany, wash it with very weak lime water, which allow to dry thoroughly. This gives a rich red color, more closely matching old mahogany than any other stain that can be used.

Antique mahogany may be obtained as follows: Take one-third linseed oil, two-thirds turpentine; coat woodwork and wipe off clean. When thoroughly dry, coat with bichromate of potash; then fill with a dark, rich filler; then shellac and give three coats of varnish and rub with pumice-stone and water, then oil and wipe off clean. If an extra good job is required, give woodwork one heavy coat of polishing varnish after being rubbed in water; then rub again in water and polish. In finishing mahogany, some put on the bichromate of potash without oiling, but they do not get as good a color. All mahogany should be oiled first, unless you want a very light color; then it should have a thin coat of shellac first.

In repolishing and reviving old work, first clean off all dust from the corners and rebates, then wipe all the polished portions with warm water and soda, and allow them to dry. Mahogany treated with spirit varnish is seldom satisfactory, but it is one of the best woods known for showing the fine effects of French polishing. Couch legs and chair turnings are generally bodied up with the brush, and finished with the rubber. If the surface is in fair condition after washing, no filling will be required; a rubber of good French polish worked out dry with spirit, and afterwards spirited out, or glazed, will give the desired result, if properly done. The polish will require staining with a little Bismarck brown or brown aniline dye, to brighten up the color. It would be a great advantage, and well worth the outlay, to put fresh gimp or leather banding round the borders; but this should not be done until the show-wood portions are repolished.

Walnut Finishing

Walnut (*Juglans Nigra*).—As this wood is seldom or never stained, it is unnecessary to say more about it other than it may be treated like oak, cherry or birch. It looks well filled and finished in shellac. Birch stained and properly finished looks very much like walnut, and with a little care in getting a proper tint in the stain, can scarcely be known from the real thing. "Filling," in walnut finishing, is one of the most important processes; if the richness of the wood is desired to be shown, as much depends on the "tint" as on the filling material.

Ordinary whitewood can be given the appearance of black walnut by first thoroughly drying the wood and then washing two or three times with a strong aqueous solution of extract of walnut peel. When nearly dry, the wood thus treated is washed over with a solution made of one part (by weight) of bichromate of potash in five parts of boiling water. After drying thoroughly, rub and polish.

A number of recipes for making and applying stains to imitate walnut are given elsewhere in this volume, which see.

Regarding Cypress

Cypress (*Cupressus sempervierens*).—The light, the dark and the bald are good woods and are coming more and more into favor every day.

This wood contains a very small amount of resin, and a very high polish can be given it; in fact, because of its not being affected by moisture, it is being used for cisterns, hogsheads, and sugar, molasses and honey barrels. The red cypress is the favorite, and some of it is so heavy that it will sink upon being placed in water. The white variety is much lighter, and will float after being deadened a short while before being cut, but it has not the firm grain of the red. The red cypress has a straight trunk with a small top, and the bark when cut has a reddish tint. These woods may be treated like cherry or birch with good results. They look well when left their natural color and finished "dead finish."

Concerning the use of cypress for inside finish, it is all right if properly dried, prepared and put in place, but dry it must be, and there will be no trouble with its staying in place or shrinking any more than any other kind of wood. It may be remarked, however, that cypress is an exceedingly hard wood to thoroughly dry, but for a low-cost material there is nothing to equal it in appearance. Get good, even-colored cypress, finish it well and some people could not tell it from red birch. If one cares to have it stained, it takes first rate. In finishing up cypress, the painters' work is the most difficult, for if the proper materials are not put on the grain is very liable to rise, which will spoil the good effect and will show even after being rubbed down. It is susceptible of a very high polish, and when finished in the natural color of the wood is very handsome. It is used by architects as a basis for the ivory white finish many people fancy, but in any event the wood when used for interiors possesses too much natural beauty to cover it with paint.

As cypress costs less than any other suitable wood for exterior work, it is not only more durable, but it will take paint better than other woods, and the paint will not peel off. We have seen buildings shingled with cypress upon both roof and walls upon which no stain or paint had been used. In time such buildings take on the beautiful gray color which is so greatly admired by many people, especially for a country or suburban residence. The natural qualities of the wood make it possible to use either shingles or clapboards in this way without paint, and there is probably no other wood upon which vines can be grown with so much safety from injurious effects.

Cypress, viewed from the standpoint of the finisher, is no less remarkable than when viewed from almost every standpoint. There is no wood which can be finished more economically, or which is more susceptible to the finer handiwork of the finisher and polisher. If the work is properly done, the result will be satisfactory in either case. It is true, notwithstanding, that the fine natural appearance of cypress is often greatly marred or even ruined by faulty methods of treatment, and for that reason care should be exercised in finishing it. The best results are obtained through the use of pure grain alcohol white shellac, which should be purchased of a thoroughly responsible dealer. Better results can be obtained from this quality of shellac than from the more expensive "refined shellac," so called.

Cypress requires no filling or sealing, and, if it is desired to permanently preserve the natural color of the wood, no oil or oily substance should be applied until the final rubbing down after the wood is well protected with shellac. We recommend three or more coats of shellac, as may be desired, each coat to be smoothed down with fine sandpaper, while the final coat may be rubbed down with pumice-stone and oil to produce a dead finish, or what is sometimes termed "egg shell" finish. The final coat may be left bright, if preferred, or after rubbing down to a dead finish it may be given a French polish, according to the usual methods.

Cypress will take stains well, but we have never favored the staining of the wood or the use of any color whatever in the finish of it, for it is far too handsome to disguise in any way.

Rosewood

Rosewood (*Dalbergia Nigra*).—It seldom falls to the lot of the ordinary finisher to have to "try his hand" on the genuine wood, but sometimes it *does happen* and it is just as well that he should be armed with the means to wrestle with the work if such is ever thrown in his way. To finish rosewood requires about the same treatment as mahogany, though, as a matter of fact, many pieces of rosewood will be found to have a coarser grain than mahogany, and will require much care in filling. The main thing to be observed is to see that the filling is a shade or two darker than the wood to be filled, before any varnish is laid on. For imitation of rosewood I give below a few recipes:

Take ½ lb. of logwood, boil it with 3 pints of water till it is of a very dark red, to which add about half an ounce of salt of tartar. When boiling hot, stain your wood with two or three coats, taking care that it is nearly dry between each; then, with a stiff, flat brush, such as is used for graining, make streaks with a very deep black stain, which if carefully executed will be very near the appearance of dark rosewood. The following is another method: Stain your wood all over with a black stain, and when dry, with a brush as above dipped in the bright liquid, form real veins in imitation of the grain of rosewood, which will produce, when well managed, a beautiful effect. A handy brush for the purpose of graining may be made by taking a flat brush, such as used for varnishing, and cutting the sharp points of the hairs and making the edge irregular; by cutting out a few hairs here and there the grain may be imitated with great accuracy.

This is suitable to pine, cedar, cypress, whitewood, basswood, while the following should only be used in mahogany, cherry, or birch: Spread on the surface of the material a concentrated solution of hypermanganate of potassa, to act until the desired shade is obtained. Five minutes suffice, ordinarily, to give a deep color, a few trials indicating the proper proportions. The hypermanganate of potassa is decomposed by the vegetable fiber, with the precipitation of brown peroxide of manganese, while the influence of the potassa, at the same time set free, fixes in a durable manner the fibers. When the action is terminated the wood is carefully washed with water, dried, and then oiled and polished in the usual manner. The effect produced by this process in several woods is really remarkable.

It has been a mystery to many people why the dark wood so highly prized for furniture is called "rosewood." Its color certainly does not look much like a rose, so we must look for some other reason. It is claimed by some that when the tree is first cut the wood possesses a very strong rose-like fragrance, hence the name. This is the most probable reason for its name. There are about a half dozen kinds of rosewood trees. The varieties are found in South America, and in the East Indies and neighboring islands. Sometimes the trees grow so large that boards or planks four feet broad and ten feet in length can be cut from them. The broad boards are used for the tops of pianofortes. When growing in the forests the tree is remarkable for its beauty, but such is its value in manufacturing as an ornamental wood, that some of the forests where it once grew abundantly now have scarcely a single specimen left.

To repolish old work, such as a rosewood piano or similar articles, the following method may be adopted:

As a rule, polished rosewood pianos are not easily kept in good condition; constant cleaning and an occasional polishing are required, especially in the case of pianos that are faced with genuine rosewood veneer, which has a coarse, open grain, and is of a somewhat oily nature. Sometimes the grain-filler that is used by the polishers will ooze out and cause an uneven surface. Plaster of Paris sometimes forms the basis of a filling, and this is apt to work out white, and becomes more apparent as the dye that has been used to enrich the color of the polish fades away through exposure to strong sunlight. It must not be forgotten that many so-called rosewood pianos are not faced with genuine rosewood veneer; the more correct term to apply to such pianos is "rosewood finish." The method by which this finish is obtained depends largely on the value of the instrument. In most cases the object of the maker is to impart a uniform color (frequently called chippendale) to the wood, and in order to obtain this end much coloring matter is used; such an excessive use of color has a tendency to destroy or imperil the nature of the polish, and accounts for much of the dullness, uneven surface, or variations of color that are more noticeable on some parts of the instrument than on other parts.

Finishing Redwood

Redwood, as a wood to hold its place after worked, has no equal. The shrinkage between green and bone-dry is very little, and after it has been ordinarily seasoned the shrinkage is very little. It does not keep growing a little narrower every year, as a white pine board sometimes does; consequently all tendency to warping and twisting is done away with.

As a finishing wood for interior house finish in the natural color it has no superior among the long list of American woods. It is, however, quite necessary that the work be properly done; the main point to be observed in finishing in natural color is to avoid the use of linseed oil, as it stains the wood a dingy color. The best finishers on this coast invariably use shellac for filler, applied rather thin, so that the wood will absorb it and thereby make a hard surface, which protects the wood from bruising, and for last coat use the best grade of shellac or hard oil.

For an Egg-Shell Gloss.—One coat of orange shellac, sandpapered to a smooth surface, followed by two or three coats of Berry Brothers' (white or light) hard oil finish; rub first coats with hair-cloth or curled hair, and the last coat with pulverized pumice-stone and raw linseed oil.

For a Dull Finish.—Specify that the last coat be rubbed with pulverized pumice-stone and water, instead of oil.

For a Polished Finish.—Specify that the last coat be rubbed first with pulverized pumice-stone and water, and then with pulverized rotten stone and water, and for a *piano finish* specify a further rubbing with Berry Brothers' furniture polish, used with a little pulverized rotten stone, applied with a piece of soft felt or flannel.

If a rubbed finish is not desired, omit the specifica‚ tions for rubbing the last coat.

White Pine Finishing

Pine (*Pinus Strobus*).—If oak is the king of woods, pine is most assuredly "president," for it is at once the most useful and the most democratic of woods. It is found in the halls of the great and powerful, and in the cottage of the most humble among us. It is strong and vigorous, plain or ornamental, and is not out of place either in the backwoodsman's cabin or in the stately cathedral, and like a true man of the world, it adapts itself to every condition that circumstances may place it in.

Pine can be made to look like any known wood, but is at its best when left natural and finished in clear shellac. There is no wood grows, that will convey so cheerful a feeling to the beholder as yellow or white pine finished in a natural state. Next to being finished in a natural state, is to imitate mahogany or light cherry, which coloring it takes readily.

Where the pine—of any kind—is to be either stained or left natural, it should be "quarter sawed," as it will show a finer grain, shrink less, and last longer. The softness of its texture and its susceptibility to injury may have had some influence in preventing its general use for ornamental purposes, but the wood can be "filled," so that much of this objection is removed.

Most of the stains given under previous heads are applicable to pine. I add, however, a few more, so that the workmen may have a number of recipes to draw from.

For Walnut.—1. Dissolve by boiling 1 part each of Epsom salt and permanganate of potash in about 25 parts of water. This stain may be improved by adding a little eosine, and it works best when applied hot.

2. Catechu broken into crumbs and boiled in about twice its bulk of water until dissolved. To darken to the required depth, add bichromate of potash previously dissolved in about eight times its equivalent of water. If the deep yellow shade peculiar to the Southern walnut be required, add yellow chromate of potash. For the reddish shade of the Northern wood, add more eosine.

3. For oil stain, use Vandyke brown toned up with the siennas, the colors being strictly pure and finely ground in oil, and diluted with turpentine and a small amount of japan.

4. Burnt Turkey umber mixed in the same way as the former.

5. Mix together, by stirring, 1 quart spirits of turpentine, 1 pint asphaltum varnish, 1 pint of japan, 1 lb. dry burnt umber, 1 lb. dry Venetian red; apply with a brush. This stain is transparent, and allows the grain of the wood to show through.

6. Boil 1½ ounces washing soda and ¼ ounce bichromate of potash, in 1 quart of water; add 2½ bunces Vandyke brown. This stain may be used either hot or cold.

7. With a brush apply a thin solution of permanganate of potassa in water, until the desired color is produced, allowing each coat to dry before another is applied.

For Mahogany or Cherry.—1. For mahogany, use a pint of turpentine and an ounce of color known as dragons' blood. Dissolve and shake well before applying. For ebony, use hot liquor from logwood chips. and after dry apply a coat of tincture of steel.

For walnut, use 2 ounces of washing soda, darkened with Vandyke brown in water. Add 2 ounces of bichromate of potash in 1½ pints of water.

2. Mix together, by stirring, 1 quart of spirits of turpentine, 1 pint of varnish, and 1 lb. of dry burnt sienna; apply with a brush, and after it has been on about five minutes wipe it off with rags. This stain takes about 12 hours to dry.

3. Take 1 quart alcohol, 2 ounces of dragons' blood; pulverize the latter along with ¼ ounce of alkanet root; mix, and let stand in a warm place a couple of days. Shake frequently in the meantime. Apply with a sponge or brush. Two or three coats may be required. This makes a fine stain.

For Rosewood.—1. Mix in a bottle ¼ lb. of extract of logwood, 1 ounce salts of tartar, and 1 pint of water; in another bottle, put 1 lb. of old iron in small pieces, and 1 pint of vinegar, which, after standing 24 hours, will be ready for use; make a hard, stiff brush with a piece of rattan sharpened at one end in a wedge shape, pounding it so as to separate the fiber. Mix in 1 pint of varnish ¼ lb. of finely-powdered rose-pink. The materials are now ready, and the first thing in the process is to stain the wood with the logwood stain; give two coats of this, allowing the first to become nearly dry before applying the second; then dip the rattan brush in the vinegar, and with it form the grain, after which give the work a coat of the varnish and rose-pink. There can be no definite directions given for graining, except to study the natural wood and imitate it as nearly as possible. With the above materials skillfully applied, any common wood can be made to resemble rosewood so nearly that it will take a good judge to distinguish the difference.

2. Boil 1 lb. of logwood in 1 gallon of water, add a double handful of walnut shell, boil the whole again, strain the liquor and add to it 1 pint of the best vinegar. It is then ready for use. Apply it boiling hot, and when the wood is dry, form red veins in imitation of the grain of rosewood with a brush dipped in the following solution: Nitric acid, 1 pint; metallic tin, 1 ounce; sal ammoniac, 1 ounce. Mix and set aside to dissolve, occasionally shaking. If carefully executed it will give the appearance of dark rosewood.

For surface stains the following are sometimes used. The colors are all to be mixed with very thin glue size, laid on warm with a soft woolen material, and the wood wiped dry after application. All the colors used in staining should be well pulverized, and before use the liquid should be strained.

Imitation Oak Stain.—Equal parts burnt umber and brown ocher.

Imitation Mahogany Stain.—One part Venetian red, and two parts yellow lead.

Imitation Rosewood Stain.—Venetian red, darkened with lampblack to required shade.

Imitation Walnut Stain.—Burnt umber and yellow ocher, mixed in proportions to give desired shade.

Before leaving the subject of pine, it may be as well to say a few words regarding the long-leaved, or Georgia pine (*Pinus Pulustris*), as a great deal of it is used now in and about the city of New York, Chicago, and other large centers. This wood is very fine, strong and lasting. Some of it is insusceptible of fine finish, but the best success with it is when treated with shellac finish. In all other respects, when used as a finishing material, it may be treated the same as ordinary pine.

The softness of white pine and its susceptibility to injury may have had some influence in preventing its general use for ornamental purposes, but the wood can be "filled," so that much of this objection is removed. Its pure white color—white as compared with other woods—recommends it for purposes for which holly has been heretofore used; and the size of the timber from which clear lumber may be cut is greatly in its favor, boards of a width of sixteen and even twenty inches being not uncommon, with no shade of distinction between sap-wood and heart, and only the faintest perceptible grain.

Some specimens lately examined show a greatly enhanced beauty by very simple treatment—the filling with warm shellac varnish, bleached shellac in alcohol, applied with a brush while warm. Several coats are given, the last coat being rubbed with pumice and rotten stone moistened with water, not oil. A finish of a flowing coat of copal varnish completes the preparation. Thus treated, the wood is of a faint creamy tint, with an appearance of semi-transparency. Beautiful gradations of tone were obtained by panels of this prepared pine, mouldings of holly, and stiles of curly or bird's-eye maple, and fine contrasts were made with the pine and oiled black walnut.

For an Egg-Shell Gloss.—One coat of shellac (white shellac if the natural color of the wood is to be preserved, or orange shellac if the wood is to be stained, or is desired to be darker in tone than the natural color), sandpaper to a smooth surface, and follow with two or three coats of Berry Brothers' (white or light) hard oil finish (specify white hard oil finish if it is desired to retain the natural color of white pine); rub first coats with hair-cloth or curled hair, and the last coat with pulverized pumice-stone and raw linseed oil.

For a Dull Finish.—Specify that the last coat be rubbed with pulverized pumice-stone and water, instead of oil.

For a Polished Finish.—Specify that the last coat be rubbed first with pulverized pumice-stone and water, and then with pulverized rotten stone and water, and for a *piano finish* specify a further rubbing with Berry Brothers' furniture polish, used with a little pulverized rotten stone, applied with a piece of soft felt or flannel.

If a rubbed finish is not desired, omit the specifications for rubbing the last coat.

One of the best ways, though perhaps not the cheapest way, to finish white pine is to see that the work is well sandpapered with the grain, then thoroighly dusted. Give it at least one coat of white shellac varnish and one coat of inside varnish. Should this prove to be too expensive, substitute liquid filler for the shellac. For hard or yellow pine finish apply one coat of orange shellac varnish and one or two coats light hard oil finish, or omit the shellac and apply hard finish instead. A filler is not required for this wood. In every instance, however, whether shellac varnish, liquid filler or hard oil finish is used, care must be taken that the first coat is thoroughly dry and hard before applying the succeeding coat, or the latter is liable to sink in, causing lack of luster.

Maple

Maple (*Acer pseudo platanus*).—This is a close-grained wood and needs no filling; it should always be finished in its own color, and that not darkened but kept as light as possible by the use of white shellac for filling and the whitest ivory varnish to be found. Most manufacturers of varnish make an article from carefully selected gums that is intended for such a use. It goes without the saying and as a matter of course that hard maple takes on the finest polish of any kind of the woods.

For an Egg-Shell Gloss.—One coat of white shellac sandpapered to a smooth surface, followed by two or three coats of Berry Brothers' or other reliable white hard oil finish; rub first coats with hair-cloth or curled hair, and the last coat with pulverized pumice-stone and raw linseed oil.

For a Dull Finish.—Specify that the last coat be rubbed with pulverized pumice-stone and water, instead of oil.

For a Polished Finish.—Specify that the last coat be rubbed first with pulverized pumice-stone and water, and then with pulverized rotten stone and water, and for a *piano finish* specify a further rubbing with furniture polish, used with a little pulverized rotten stone, applied with a piece of soft felt or flannel.

If a rubbed finish is not desired, omit the specifications for rubbing the last coat.

White and Black Ash

Ash (*Fraxinus excelsior*).—This wood is now used very much by cabinetmakers and house-joiners in place of oak, and I have often seen furniture palmed off to unsuspecting customers as antique oak, and in one instance I knew of an architect who specified oak, and who "passed" a mixture of white and black ash as oak, either knowingly or otherwise. I am not sure that the owners in either case lost anything, **for**

good sound Canadian ash is decidedly better than dosey red oak.

In finishing ash, either black or white, the same methods are adopted as for finishing oak, and similar processes will give similar results. Ingenious stainers and finishers can make ash resemble oak wainscot, in vein and color, so correctly that it is almost impossible for the most experienced connoisseur to distinguish the genuine from the spurious. In order to do this some finishers make a commencement by sketching out, upon certain parts of the ash exterior, the requisite white veins, by means of a camel-hair pencil, with white stain; that done, they coat the veins with thin varnish, and then darken the general ground, dealing carefully throughout the entire process with the veined portions. Others stain and embody, i.e., French polish, the ash with the ordinary preparation, after which they pursue an operative course termed "hamping"; that is, scratching fancifully, so as to form the veins, upon different parts of the coated surface, before it gets time to harden, with a saturated rag. The former process is, however, the more suitable of the two.

For an Egg-Shell Gloss.—One coat of filler to match the color of the wood, followed by one coat of white shellac sandpapered to a smooth surface, and two or three coats of white or light hard oil finish; rub first coats with hair-cloth or curled hair, and the last coat with pulverized pumice-stone and raw linseed oil.

For a Dull Finish.—Specify that the last coat be rubbed with pulverized pumice-stone and water, instead of oil.

For a Polished Finish.—Specify that the last coat be rubbed first with pulverized pumice-stone and water and then with pulverized rotten stone and water, and for a *piano finish* specify a further rubbing with Berry Brothers' furniture polish, used with a little pulverized rotten stone, applied with a piece of soft felt or flannel.

If a rubbed finish is not desired, omit the specifications for rubbing the last coat.

Other Woods

Cedar, White (*Cupressus thuyoides*), which is really a spruce, and all similar woods, should never be finished in a natural state. Deep stains or surface stains should always be employed on these woods if they are not to be painted.

Beech (*Fagus ferruginea*).—This is one of the unnoticed woods of former years, but is now gradually gaining in favor as a decorative wood. It is cheap and also quite abundant, while the more popular hardwoods are beginning to grow scarcer and higher in price. Beechwood has a fine grain, is quite durable, and can be used in the manufacture of furniture and for decorative purposes generally. The red variety has a handsome appearance and is especially suitable for use where a good imitation of cherry is desired.

If "quarter sawed" it shows a fine grain and has a character distinctly its own which I think has never been properly appreciated. When quartered, properly finished, filled and polished, it looks something like dark leopard wood. It will assume a dark mahogany color if prepared like cherry or birch, or it may be made to appear like walnut if treated with walnut stains and finish.

Elm (*Ulmus Americana*), **Chestnut** (*Castanea vesca*), **Butternut** (*Juglans cinerea*).—These three woods are often used in inferior work, and are very soft and easily dented. The best is perhaps the elm, which does very well for bath-room finish, panels for ash doors and similar work. All require a great deal of "filling," and this should be well rubbed in if a good job is required. All of these woods have a very coarse grain, but if care is taken in selecting the material, very odd and sometimes pleasing effects may be obtained. Any of the stains used on pine will answer for these woods, dependent, of course, on the tints desired. The best result with these woods is derived by giving the work one coat of shellac after filling and staining; then sandpaper well and apply your varnish or oil finish or whatever you purpose finishing in.

Sycamore, or Buttonwood, as it is sometimes called (*acer pseudo platanus*), when quarter sawed and properly finished makes a good appearance, and in many cases is superseding cherry owing to its beauty and cheapness. Heretofore its natural beauty has been destroyed in many cases by staining the wood, and thus preventing the development of many chemical changes which take place and are thrown to the surface when properly treated. When quarter sawed, a light-bodied and light-colored shellac should be used, when by a natural chemical process a beautiful silver leaf is developed and the surface assumes a charming pink hue.

Hemlock (*Abies Canadensis*).—This is rarely used for finishing, owing to its brittleness and splintery nature. Clean boards, however, made into panels or other similar work and then finished in natural color, assume a very handsome appearance. It has a pink tint of the most delicate kind, which improves and mellows with age. It should be finished in the same manner as pine.

I have purposely left unmentioned a number of our finest woods, concerning which I should have more to say, would it not increase the size of this volume beyond reasonable dimensions. The workmen will know, from what has already been said, how these light woods may be treated, and I may add that the various maples, though rather light in color for general purposes, make the most delicate of finish. Gum-wood of late has been used to some extent and is not hard to deal with when to be finished. It should be treated about as cherry and birch, and finished in a similar way, and when done nicely the work looks quite well.

Redwood is getting to be a favorite wood with some builders, owing no doubt to its cheapness, and its easiness to work. It may be classed among the soft woods and requires treating about the same as chestnut, butternut and the cedars. The following formula and directions have been highly recommended as a finish, but for my own part I prefer using Wheeler's filler, as I think better results are obtained. Here is the formula:

Take 1 quart spirits turpentine.

Add 1 pound corn-starch.

Add ¼ pound burnt sienna.

Add 1 tablespoonful raw linseed oil.

Add 1 tablespoonful brown japan.

Mix thoroughly, apply with a brush, let it stand, say fifteen minutes; rub off all you can with fine shavings or a soft rag, then let it stand *at least twenty-four hours*, that it may sink into and *harden* the fibers of the wood, afterward apply two coats of white shellac; rub down well with fine flint paper, then put on from two to five coats best polishing varnish; after it is well dried, rub with water and pumice-stone ground very fine, stand a day to dry; after being washed, clean with chamois, rub with water and rotten stone; dry, wash clean, as before, and rub with olive oil until dry.

Some use cork for sandpapering and polishing, but a smooth block of hardwood like maple is better. When treated in this way redwood will be found the peer of any wood for real beauty and life as a house trim or finish. By proper attention, redwood may be made to appear like mahogany or cherry, but its softness is very much against it.

SOME GENERAL RECIPES

Ebonizing.—Prepare some shellac varnish by dissolving half a pound of gum shellac in a quart of alcohol; put in a tightly corked bottle, set in a warm place, and shake frequently. When completely dissolved, add sufficient dry lamp or bone black. It will require only about a heaping tablespoonful to half a pint of varnish. Should it be too thick to flow easily from the brush and spread evenly, add alcohol. Give the wood two or three coats of this, which can be done within two days. For fine work give three coats of the black varnish and allow it to dry thoroughly; then take some oo or ooo sandpaper and rub the work down lightly until the surface is smooth and even, being careful not to rub through to the wood. Then apply two or three coats more, and rub down again with the sandpaper; give two coats more and allow it to dry perfectly hard. Make some rubbers of felt—an old felt hat is good—by tacking the felt on blocks of soft wood of proper shape, which should be flat, concave or convex to fit the work. Apply a few drops of sewing machine oil to the felt and sprinkle on a pinch of pulverized pumice-stone. Rub the work with this, keeping the rubber moist with oil and supplied with the pumice, until the wood shows a perfectly smooth surface, without any gloss. When it has all been gone over, wipe off the surface and rub clean with dry flannel cloths. The result is a fine imitation of ebony. If the shellac varnish is colored with aniline instead of lampblack, the finished surface is smoother and finer.

Others. — 1. Infuse gall-nut in vinegar, in which rusty nails have been soaked; paint the wood with this, polish and burnish when dry. 2. Wash the wood repeatedly with a solution of sulphate of iron, made by dissolving 2 ounces of sulphate in a pint of hot water. When dry, apply a hot decoction of logwood and nut-galls two or three times. When dry, clean with a wet sponge and then polish. 3. Brush the wood with a strong decoction of logwood chips several times. When dry, give it a coat of vinegar in which rusty iron has been placed. Dissolve beeswax in turpentine by setting in a warm place; apply warm with a brush, and rub it till it shines. 4. Wash with a concentrated aqueous solution of logwood several times, and then with a solution of acetate of iron 40° Baume. Repeat till a deep black is produced. 5. Put 2 ounces of logwood chips with 1¼ ounces of copperas in a quart of water, boil, and lay on hot. When dry, wet the surface again with 2 ounces of steel filings dissolved in half a pint of vinegar. When dry again, sandpaper smooth, then oil, then fill it with powdered drop black mixed in the filler. Work to be ebonized should be smooth and free from holes. Give it a light coat of quick-drying varnish, then rub with finely pulverized pumice-stone and linseed oil until very smooth. 6. Boil ½ pound of chip logwood in 2 quarts of water, and add ½ ounce of verdigris and ½ ounce of copperas, strain, and put in ½ pound of rusty steel filings. With this go over the work a second time. 7. A pound of logwood boiled in 4 quarts of water, add 2 handfuls of walnut shells or peel, boil up again, take out the chips, add a pint of vinegar, and apply boiling. Afterwards dissolve 1 ounce of green copperas in a quart of boiling water and apply hot. 8. First sponge the wood with a solution of chlorhydrate of aniline in water, to which a small quantity of copper chloride is added. When dry, go over again with a solution of potassium bichromate. Repeat this twice or thrice. 9. One gallon of vinegar, ½ pound of green copperas, ¼ pound of China blue, 2 ounces nut-gall, 2 pounds extract of logwood. Boil all these over a slow fire, and add half a pint of iron-rust. Apply as usual. A good varnish for ebonized work is made by dissolving in alcohol some black wax.

STAINS

Many excellent stains for pine may be obtained by using the ordinary graining colors, Vandyke brown, raw and burnt sienna, ultramarine blue, etc., applied with a brush, without previous preparation, and then wiped off with a cloth—a method that brings out clearly the grain or marks of the wood, which in pitch pine, now being extensively used for fittings, are often extremely beautiful. A better method for general work—French polish being ordinarily too expensive—is, where dark oak or mahogany stains are not wanted, light varnishes, of which two coats are to be applied. The glue size with which the work is first coated, in order to fill up the pores of the wood, should not be too thick, as in that case it is liable to crack.

Logwood, lime, brown soft-soap, dyed oil, sulphate of iron, nitrate of silver exposed to the sun's rays, carbonate of soda, bichromate and permanganate of potash, and other alkaline preparations, are used for darkening the wood; the last three are specially recommended. The solution is applied by dissolving 1 ounce of the alkali in 2 gills of boiling water, diluted to the required tone. The surface is saturated with a sponge or flannel, and immediately dried with soft rags. The carbonate is used for dark woods. Oil tinged with rose madder may be applied to hardwoods like birch, and a red oil is prepared from soaked alkanet root in linseed oil. The grain of yellow pine can be brought out by two or three coats of japan, much diluted with turpentine, and afterwards oiled and rubbed. To give mahogany the appearance of age, lime water used before oiling is a good plan. In staining wood, the best and most transparent effect is obtained by repeated light coats of the same. For oak stain a strong solution of oxalic acid is employed; for mahogany, dilute nitric acid. A primary coat or a coat of wood fillers is advantageous. For mahogany stains, the following are given: 2 ounces of dragons' blood dissolved in 1 quart of rectified spirits of wine, well shaken, or raw sienna in beer, with burnt sienna to give the required tone; for darker stains boil ½ pound of madder and 2 ounces of logwood chips in 1 gallon of water, and brush the decoction while hot over the wood; when dry, paint with a solution of 2 ounces of potash in 1 quart of water. A solution of permanganate of potash forms a rapid and excellent brown stain.

Oak or ash may be stained brown by using linseed
oil and benzine half and half, and burnt umber or
Vandyke brown incorporated with this. Maple can
be stained green-gray by using copperas in water; oak
will also be changed to a dark green-blue through the
same agency, the effect on ash being various shades of
olive green. Ammonia applied to oak produces the
bronze olive tint now used so much by architects.

Wash any compact wood with a boiling decoction of
logwood three or four times, allowing it to dry between
each application. Then wash it with a solution of
acetate of iron, which is made by dissolving iron filings
in vinegar. This stain is very black, and penetrates
to a considerable depth into the wood, so that ordinary
scratching or chipping does not show the original
color.

A wash of 1 part of nitric acid in 10 parts of water
will, if well done, impart a stain resembling mahogany
to pine wood which does not contain much resin.
When the wood is thoroughly dry, shellac varnish will
impart a fine finish to the surface. A glaze of carmine
or lake will produce a rosewood finish. A turpentine
extract of alkanet root produces a beautiful stain which
admits of French polishing. Asphaltum, thinned with
turpentine, makes an excellent mahogany color on
new wood.

A Method of imparting to any plain white wood
the appearance of cedar wood is as follows: Mix 2
parts of catechu, and 1 part of caustic soda, in 100
parts of water (all by weight). The article to be
stained should be boiled in this solution for some
hours, and is then rinsed in clear water and dried. If
the desired depth of tone has not been obtained, a
second boiling must be resorted to. This stain is said

to sink so deep into the wood that even thick sheets of veneer thus treated will be colored right through; while other wood articles thus stained may be safely manipulated without any fear of the original color of the wood showing through.

For a Brown Stain.—Boil 1 pound of the brown pigment called Terre de Cassel with 4 quarts of water, until it is reduced one-third. Mix 2 ounces (Troy) of white potash with sufficient water to dissolve it, and mix with the Terre de Cassel. This stain must be applied with a brush, two or even three times, according to the depth of the shade required.

Another.—Paint the wood with a solution made by boiling one pot of catechu with 30 parts of water and a little soda; when dry, paint over with another solution made of 1 part bichromate of potash and 30 parts water. By a little difference in the mode of treatment, and by varying the strength of the solutions, several shades of color may be given.

Staining Carved Panels.—First, oil the panels with linseed oil, then mix a little powdered water stain (this is best mixed with rain water), and brush over the stain with a hog's hair brush, using as little stain as possible. When dry, give a coat of button polish, which must be laid on with a camel-hair brush. Care must be taken not to form puddles in the carvings, either with the stain or polish. When the polish is quite dry, the panels will be improved by rubbing with a piece of clean linen rag. By following the foregoing, a good effect is obtained, and, by oiling before staining, the fibers of the wood will not rise.

Staining Spirit Varnish for Furniture, etc.—White hard spirit varnish, as it comes from the makers, is generally much too thick for mixing with stain, and does not distribute evenly. Stained varnishes should be diluted with spirits or turpentine, whichever may be the solvent. Two applications of thin varnish will give far more satisfactory results than one heavy coating of thick varnish. Brown hard spirit varnish with the addition of red stain will give much better results. When the work is streaky, it would be advisable to remove it by softening with spirits or grinding down with pumice-stone powder and water, then using varnish as advised above. Oil varnish, such as church oak varnish, is generally considered too garish for use on furniture. In repolishing old work, the idea of filling up a deep scratch with beeswax and resin is a good one, but it requires considerable practice to get a quite level surface and a perfect match as regards color. The surface that has been disturbed during the operation may be treated by coloring up; that is, bodying up to fill the grain of the wood; then, by blending together suitable colors, mixed in 1 part of polish to 3 parts of spirits, carefully penciling in till a perfect match is secured.

DYEING WOODS

For a Black Dye.—Put 6 pounds of chip logwood into the copper, with as many veneers as it will conveniently hold, without pressing too tight; fill it with water, and let it boil *slowly* for about three hours; then add ½ pound of powdered verdigris, ½ pound of copperas, and 4 ounces of bruised nut-galls; fill the copper up with vinegar as the water evaporates; let it boil gently two hours each day till the wood is dyed through.

Another.—Procure some liquor from a tanner's pit, or make a decoction of oak bark, and to every gallon of the liquor add a quarter of a pound of green copperas, and mix them well together; put the liquor into the copper, and make it quite hot, but not boil; immerse the veneers in it, and let them remain for an hour; take them out, and expose them to the air, till it has penetrated its substance; then add some logwood to the solution, place the veneers again in it, and let it simmer for two or three hours; let the whole cool gradually, dry the veneers in the shade, and they will be a very fine black.

Dyeing wood is mostly applied for the purpose of veneers, while staining is more generally had recourse to, to give the desired color to the article after it has been manufactured. In the one case the color should penetrate throughout, while in the latter the surface is all that is essential.

In dyeing pear tree, holly and beech, take the best black; but for most colors holly is preferable. It is also best to have wood as young and as newly cut as possible. After the veneers are cut, they should be allowed to lie in a trough of water for four or five days before they are put into the copper, as the water, acting as a purgative to the wood, brings out an abundance of slimy matter, which must be removed, or the wood will never be a good color. After this purificatory process, they should be dried in the open air for at least twelve hours. They are then ready for the copper. By these simple means the color will strike much quicker, and be of a brighter hue. It would also add to the improvement of the colors, if, after the veneers have boiled a few hours, they are taken out, dried in the air, and again immersed in the coloring copper. Always dry veneers in the open air, for fire invariably injures the colors.

Fine Blue Dye.—Into a clean glass bottle put 1 pound of oil of vitriol, and 4 ounces of the best indigo pounded in a mortar (take care to set the bottle in a basin or earthen glazed pan, as it will ferment); then put the veneers into a copper or stone trough; fill it rather more than one-third with water, and add as much of the vitriol and indigo (stirring it about) as will make a fine blue, which may be known by trying it with a piece of white paper or wood. Let the veneers remain till the dye has struck through.

The color will be much improved if the solution of indigo in vitriol be kept a few weeks before using it. The color will strike better if the veneers are boiled in plain water till completely soaked through, and then allowed for a few hours to dry partially, previous to being immersed in the dye.

Another.—Throw pieces of quicklime into soft water; stir it well; when settled, strain or pour off the clear part; then to every gallon add ten or twelve ounces of the best turnsole; put the whole into the copper with the veneers, which should be of white holly, and prepared as usual by boiling in water; let them simmer gently till the color has sufficiently penetrated, but be careful not to let them boil in it, as it would injure the color.

A Fine Yellow Dye.—Reduce 4 pounds of the root of barberry, by sawing, to dust, which put in a copper or brass trough; add 4 ounces of turmeric and 4 gallons of water, then put in as many white holly veneers as the liquor will cover; boil them together for three hours, often turning them; when cool, add 2 ounces of aqua fortis and the dye will strike through much sooner.

A Bright Yellow Dye.—To every gallon of water necessary to cover the veneers, add 1 pound of French berries; boil the veneers till the color has penetrated through; add to the infusion of the French berries, the liquid for brightening colors given as stated, and let the veneers remain for two or three hours, and the color will be very bright.

Bright Green Dye.—Proceed as in either of the previous recipes to produce a yellow; but instead of adding aqua fortis or the brightening liquid, add as much vitriolated indigo as will produce the desired color.

Green Dye.—Dissolve 4 ounces of the best verdigris, and of sap-green and indigo ½ ounce each, in 3 pints of the best vinegar; put in the veneers, and gently boil till the color has penetrated sufficiently.

The hue of the green may be varied by altering the proportion of the ingredients; and it is advised, unless wanted for a particular purpose, to leave out the sap-green, as it is a vegetable color very apt to change, or turn brown, when exposed to the air.

Bright Red Dye.—To 2 pounds of genuine Brazil dust, add 4 gallons of water; put in as many veneers as the liquor will cover; boil them for three hours; then add 2 ounces of alum, and 2 ounces of aqua fortis, and keep it lukewarm until it has struck through.

Red Dye.—To every pound of logwood chips, add 2 gallons of water; put in the veneers, and boil as in the last; then add a sufficient quantity of the brightening liquid till you see the color to your mind; keep the whole as warm as the finger can be borne in it, till the color has sufficiently penetrated.

The logwood chips should be picked from all foreign substances, with which it generally abounds, as bark, dirt, etc.; and it is always best when fresh cut, which may be known by its appearing of a bright-red color; for if stale, it will look brown, and not yield so much coloring matter.

Purple Dye.—To 2 pounds of chip logwood and ½ pound of Brazil dust, add 4 gallons of water, and after putting in the veneers, boil them for at least three hours; then add 6 ounces of pearlash and 2 ounces of alum; let them boil for two or three hours every day, till the color has struck through.

The Brazil dust only contributes to make the purple of a more red cast; you may, therefore, omit it, if you require a deep bluish purple.

Another. — Boil 2 pounds of logwood, either in chips or powder, in 4 gallons of water, with the veneers; after boiling till the color is well struck in, add by degrees vitriolated indigo till the purple is of the shade required, which may be known by trying it with a piece of paper· let it then boil for one hour, and keep the liquid in a milk-warm state till the color has penetrated the veneer. This method, when properly managed, will produce a brilliant purple, not so likely to fade as the foregoing.

Liquid for Brightening and Setting Colors.—To every pint of strong aqua fortis, add 1 ounce of grain tin, and a piece of sal-ammoniac of the size of a walnut; set it by to dissolve, shake the bottle round with the cork out, from time to time; in the course of two or three days it will be fit for use. This will be found an admirable liquid to add to any color, as it not only brightens it, but renders it less likely to fade from exposure to the air.

Orange Dye.—Let the veneers be dyed by either of the methods given as above, of a fine deep yellow, and while they are still wet and saturated with the dye, transfer them to the bright red dye, till the color penetrates equally throughout.

Silver Gray Dye.—Expose to the weather, in a cast-iron pot of six or eight gallons, old iron nails, hoops, etc., till covered with rust; add 1 gallon of vinegar and 2 of water, boil all well for an hour; have the veneers ready, which must be hardwood (not too dry), put them in the copper used to dye black, and pour the iron liquor over them; add 1 pound of chip logwood and 2 ounces of bruised nut-galls; then boil up another pot of the iron liquor to supply the copper with, keeping the veneers covered, and boiling two hours a day, till the required color is obtained.

Gray Dye.—Expose any quantity of old iron, or what is better, the borings of gun-barrels, etc., in any convenient vessel, and from time to time sprinkle them with spirits of salt (muriatic acid) diluted in four times its quantity of water, till they are very thickly covered with rust; then to every 6 pounds add 1 gallon of water, in which has been dissolved 2 ounces of salt of tartar; lay the veneers in the copper, and cover them with this liquid; let it boil for two or three hours till well soaked, then to every gallon of liquor add a quarter of a pound of green copperas, and keep the whole at a moderate temperature till the dye has sufficiently penetrated.

GILDING, SILVERING, BRONZING, AND COMPOSITION

In gilding, the design should be simple and fairly open, so that the gold leaf can be laid in water without too many difficulties. As the particular method of gilding here described requires the whole surface to be burnished, it will be found that a design should be treated in a broad way, as, if the surface is too much broken up, it will, on account of its being burnished, and therefore reflecting light in a high degree, be worrying in its effects, as well as entailing an enormous amount of time to lay the gold leaf in water. The clever craftsman is he who gets his effects by means as simple as possible. That piece of work will be the most satisfactory that fulfills the purpose for which it was designed, however simple the means, so long as the result is lasting.

In the design Fig. 115 now given, the frame is to take a mirror, the sight size of which should be 16⅛ inches high by 13⅛ inches wide, the full size of the plate being 17 inches by 14 inches. The extreme dimensions of the frame are 24¾ inches high, 25⅜ inches wide. The

FIG. 115

plinth will require a piece 20⅛ inches by 3⅞ inches by ⅞ inches thick; the moulding for it will require a piece ¾ inch by ⅜ inch. The pilasters are 3¼ inches by ¾ inch thick. They are tenoned into the plinth, the mortises in which should not be more than 1¼ inches deep, so that they do not come below the marginal line of the carving along it. The top rail, which shows ⅞ inch on the face of the frame, is tenoned into the pilasters; the top of the carving on the pilasters should come on a line with the rail, so as not to weaken

Fig. 116

or cut through the mortises. The hollow moulding is got out of $3\frac{1}{8}$ inches by $\frac{7}{8}$ inch, and is returned on the sides, as is the moulding on the plinth. The ogee moulding requires a piece $25\frac{3}{8}$ inches by $3\frac{5}{8}$ inches by $\frac{3}{4}$ inch; the returns in this are shaped out of the solid at each end of the piece, and it is screwed on the frame from the top.

The design for the pilasters, Fig. 116, is one-fifth full size. It is very important when carving to treat the ground freely; the worker should aim at having the ground somewhat uneven, not in an abrupt and obvious way, but with a soft up and down sort of feeling, which will, when the work is gilded, make it look ever so much more human. The result will be more interesting than if the ground is cut perfectly flat and even. The edges round the carved parts should not be set in, but should slope underneath the forms in an easy hollow. The pilasters should be grounded out a $\frac{1}{4}$ inch at the base, and more shallow at the top, so that the smaller leaves and flower should be in fainter relief. The plinth should not be grounded out deeper than $\frac{1}{8}$ inch. Pine or whitewood may be used to make the frame.

Having made the frame and put it together, we now want to gild it. The way here described is a modification of the old Italian method. First of all, it will be necessary to slake some fine plaster of Paris in water for three weeks (about 1 pound of plaster to 4 gallons of water). It should be stirred frequently the first day to prevent setting. At the end of three weeks the water is poured out, the plaster is put into a cloth, and the water squeezed out; it is then left to dry into a solid cake. When wanted for use, break a piece that will go in a small pipkin, and cover it with water for half an hour. While the plaster is soaking the frame should be got ready. Clear parchment size will now be wanted; it can be had from gilders' material dealers. Some should be melted in a pipkin, care being taken that it does not boil; if it does, its setting property will be destroyed. With a round hog-hair brush give the frame two coats of clear size, using it its full strength; then pour the water off the plaster and pour some warm size, full strength, over it and mix together; it should be mixed to the consistency of cream; warm it up, and strain through muslin, working it through with a brush; warm it again, and give a coat to the carved work with the hog-hair brush, and to the plain parts with a camel-hair mop. Four coats should be given on the carved parts, using the soft brush after the first coat, and six coats given on the plain parts. They should be given as soon as the preceding coat is set, but not dry.

When the coats are quite dry, which should be the next day, the plaster has to be smoothed down. It will be more possible to preserve the shape in the small work if emery cloth is used. When all roughness has been smoothed down, all parts should carefully be gone over with finest sandpaper.

The work has now to be prepared so that the gold

will stand burnishing. Armenian bole, which can be obtained of chemists, will be required, and some pipeclay. Blanc's will do, as it is fine. Three-fifths of bole to two of clay are ground together on a piece of glass with a muller, using some water while grinding. The grinding should proceed for a good half hour, in fact, it cannot be too well ground, as, if it is in the least gritty, it will rub through when burnishing. For the frame in hand, about half a pipkinful will be required.

Melt down some size, and put some water in a pipkin about a quarter full; add some melted size, so that there should be just a bit more size than water. Then put in the ground bole and clay, and mix well together. The mixture should flow easily from the brush, but should not be too thin. Before using all over the frame, it will be as well to try some on the back edge; allow it to dry. If it dries the same color as when put on wet, there is too much size, and it should be weakened with water. If it dries a lighter color, rub it down with D emery paper, wearing down the sharpness of the paper first, as it is too strong if used fresh. If, on being papered, it polishes without rubbing off, it is the proper strength, and may be used. Should it rub off, it is weak, and more size should be added. Having our burnish gold size the proper strength, we give the work four coats all over, using a camel-hair mop, allowing each coat to set, and taking care to take up all pools that may collect in the bottoms. When quite dry, it must be rubbed down all over with D emery paper, and then well bristled with a short hog-hair brush until there is a good polish everywhere. If there be any dull parts, there is something wrong, and the gilding should not proceed until put right, as failure will be the result. The most essential points are that the bole be well ground, and that the size is not too strong.

The frame is then ready for gilding. The water to gild with must have three teaspoonfuls of size to a small pipkin full of water. The gold is laid on in the usual way. The writer will attempt to describe how to lay it; but it is an operation that really requires demonstration. A cushion, a knife, a whole and a three tip (the whole tip to take a whole leaf, the three-tip to take smaller pieces of the leaf), a camel-hair pencil and liner; these should be placed on the same handle—one at each end will be wanted. The plinth had better be laid first, then the pilasters, all the smaller fillets and mouldings next, leaving the hollow and ogee mouldings to be laid last, as the beginner will by then have acquired some facility in the handling of the leaf. It will be better for a beginner to blow out one leaf at a time on to his cushion; it will save waste until he is more expert in the handling. To take out a leaf, hold the book almost perpendicularly, expose a gold leaf, and gently blow it on to the cushion; then take the cushion in the left hand, passing the thumb through the leaf which is under-neath; pass the knife very carefully underneath the leaf, lift it up, and spread it open on the cushion. If no corners are folded underneath it, blow gently in the center of it, so that it will be flattened out; then cut the leaf by drawing the knife firmly through the middle of it; then divide the two halves into three pieces, so that the leaf is divided into six pieces 1½ inches by 1 inch. A piece of gold is picked up by the three-tip being placed about halfway on the piece; if it does not adhere to the tip the worker should rub the tip briskly on his hair—a slight oiliness will make

the gold take kindly to the tip. Great care must be
taken to keep the knife clean and free from grease.
The tip with the piece of gold attached to it must be
held between the first two fingers of the left hand, and
the knife by the next two fingers. Take up the pencil,
wet it in the water and size, and wet the ground
between two of the forms on the plinth. The place to
be gilded must be wet, but pools must be avoided, else
the gold will not reach the ground, or, if pressed
down, the water will burst through the gold and cause
a stain. Holding the pencil and liner in the right
hand, the tip is now taken with the right hand, held
over the place where the gold has to go, and the piece
gently pressed down on it. With the liner the gold is
pressed home, and if the piece is somewhat larger than
the ground, the overlapping gold is turned into the
corners of the ground and ornament with the liner, or
faulter as gilders call it. All the ground should be
gone over in this way; gentleness is required so as not
to break the gold, and smaller pieces should be cut and
used where needed. When all the ground has been
laid over, the leaf is cut in suitable pieces to cover the
forms of the ornament. Great care must be taken in
wetting the former that the pencil does not touch the
parts already gilded, for that will mean a stain. In
laying the fillets and smaller mouldings the gold
should be used the length of the leaf, but cut the width
required to cover the space. It will be better to lay
all the narrowest parts first, going on to the wider and
gaining experience, so as to be able to use a whole
leaf on the hollow moulding. In taking up a whole
leaf with the large tip, the tip must not quite take up
half of the leaf, so as to be able to lay it without
break in the hollow. For wetting the larger parts, the

camel-hair mop should be used. The parts laid should be burnished within two or three hours of the gold being laid. With an agate or flint burnisher sound the part laid first—if it sounds dull, it is not dry enough; if sharp, it is dry and should be burnished at once. The burnisher must be used lightly and across the form. If everything has been done satisfactorily, the burnisher will go very smoothly and softly. The resultant burnish will be the color of the gold, and not dark, as, indeed, one may say black, as all ordinary burnish usually is. When the work has been burnished, the beginner will, no doubt, feel distressed at seeing parts where the gold has not adhered, but that happens to all gilders, more or less, depending on the worker's skill, and all those parts must be faulted, using a smaller pencil to wet the faults. When done, the frame should sparkle with light and shade, full of life and having a depth of color that is absent in the frames gilded in the ordinary way.

Gilding Furniture.—Gold leaf, applied to articles of furniture as a means of decoration, is used in two ways; it is applied over an ordinary varnish or other finish, in which case but little special preparation is necessary; or, as when used for picture frames, cornices, etc., it is applied to a specially prepared foundation, the basis of which is whiting, mixed with various other ingredients suggested by experience or fancy. In either case, the gold leaf is caused to adhere to the work by size specially prepared for the purpose, recipes for which follow herewith; the size being first applied to the work, and when it has become of the right consistency, the gold is laid upon it. *Oil gilding* and *burnish gilding* are different methods used to obtain certain desired effeets, the former principally for articles exposed to the weather, and for heightening the effect of incised carving or engraving, and the latter for picture frames and articles having a specially prepared foundation, whose entire surface is to be gilded. It is intended that the gold shall adhere to the work only in the places to which the size has been applied, but the smallest portion of oil or even a slight dampness may cause the gold to partially adhere to the adjoining surface, resulting in slightly ragged edges; to prevent this, before applying the size to the desired design, the entire surface is covered with a thin film of some substance perfectly free from moisture, and easily removable by water, after completion of the process.

The Requisites.—First, a sufficient quantity of leaf gold, which is of two sorts, the deep gold, as it is called, and the pale gold. The former is the best; the latter very useful, and may occasionally be introduced for variety or effect.

Second, a gilder's cushion: an oblong piece of wood, covered with rough calfskin, stuffed with flannel several times doubled, with a border of parchment, about four inches deep, at one end, to prevent the air blowing the leaves about when placed on the cushion.

Thirdly, a gilding knife, with a straight and very smooth edge, to cut the gold.

Fourthly, several camel-hair pencils in sizes, and tips, made of a few long camel's hairs put between two cards, in the same manner as hairs are put into tin cases for brushes, thus making a flat brush with a very few hairs.

Lastly, a burnisher, which is a crooked piece of agate set in a long wooden handle.

Sizes.—These are of two kinds: oil sizes are those which, when applied, present an adhesive surface, requiring the immediate laying of the gold leaf upon it; of this class is the oil size commonly used in decorating furniture; water sizes are those that are allowed to become dry and hard when applied, and are rendered adhesive when the gold is to be laid, by brushing over with water; for burnish gilding these are always employed, as oil size does not dry sufficiently hard to permit of burnishing.

Oil Size for Oil Gilding.—Grind calcined red ocher with the best and oldest drying oil. When desired for use, add sufficient oil of turpentine to make it work freely.

Parchment Size.—For preparing frames, etc. To half a pound of parchment shavings, or cuttings of white leather, add three quarts of water, and boil it in a proper vessel till reduced to nearly half the quantity; then take it off the fire, and strain it through a sieve. Be careful in the boiling to keep it well stirred, and do not let burn.

Gold Size for Burnish Gilding.—Grind fine sal-ammoniac well with a muller and stone; scrape into it a little beef suet, and grind all wet together; after which, mix in with a pallet knife a small proportion of parchment size with a double proportion of water. When about to use, add parchment size until it will just flow from the brush.

Another.—Grind a lump of tobacco pipe clay into a very stiff paste with thin size; add a small quantity of ruddle and fine black lead, ground very fine, and temper the whole with a small piece of tallow. When ready to use, reduce with parchment size until it will just flow from the brush.

Another. — Grind separately in water, 1 pound Armenian bole, 2 ounces red lead, a sufficient quantity of black lead; mix, and regrind with a small quantity of olive oil. Reduce with parchment size to the proper consistency.

To Prevent Gold Adhering.—Either one of the following methods will prevent gold leaf or bronze from adhering to the surface beyond the outlines of the sizing laid on to receive it:

1. Whiting used dry, and applied by means of a pounce bag.

2. Whiting mixed in water, and applied with a soft brush. When the water has evaporated, dust off the superfluous whiting with an ordinary paint duster. By this method a very thin coating of whiting remains, which is free from any grittiness. One advantage gained by the use of whiting thus applied is, it furnishes a whitish ground over which clear varnish or oil size may be distinctly seen as the striping progresses. After the leaf or bronze has been applied, the work must be carefully washed, so as to insure the removal of the whiting.

3. White of egg reduced with water, and applied with a piece of sponge.

4. A thin wash of starch water, either brushed on with a flat camel-hair brush, or applied with a soft sponge.

5. Take ball liquorice and water, a weak solution, and apply with a soft brush. This may be kept in a bottle ready for use at any time.

6. Cut a new potato in two, and rub over the part to be sized with the raw face exposed, allowing the juice to remain until dry.

It will be observed that any substance which interposes a film over the varnish, itself being free from tackiness and readily removed by water, will answer the purpose.

Oil Gilding.—Applying the gold.—If the wood to be gilded is finished with varnish or otherwise, no additional foundation is necessary upon which to lay the gold leaf; if the wood is not finished, after it has been smoothed and dusted, give it one or two coats of parchment size, after it is perfectly dry and hard, again smoothing the surface with fine sandpaper. That the gold may not adhere to any part of the work except where the size is hard, powder the surface lightly with whiting from a pounce bag, which is a small bag made of material sufficiently loose to permit the powdered whiting to sift through as fine dust; if preferred, any of the preceding recipes for that purpose can be used instead. Remove the surplus whiting with the dusting brush, and the work is then ready for the size. Apply this with a sable or fit brush of the proper size, carefully observing to make the outer lines of the design clear and sharp, that the work may not appear ragged. Let the size remain until it feels tacky, when the gold may be applied. This is the most difficult part of the operation, and experience is necessary before gold leaf can be laid smoothly without a wrinkle or a break. Turn a leaf of gold out of the book upon the cushion; breathe gently upon the center of the leaf and it will lie flat on the cushion; cut it to the proper size by bringing the knife perpendicularly over it, and sawing it gently until divided. Take your tip (a brush used for the purpose) and after drawing it lightly over your hair to remove any particles or dust that may be upon it, breathe upon it gently, which will dampen it sufficiently to cause the leaf of gold to adhere to it; lay the tip upon the leaf of gold and carefully transfer it to the work; blow upon it gently and it will straighten out and adhere. It may be rendered quite smooth by slightly dabbing it with a bit of cotton. In about an hour wash off the superfluous gold from the edges, with a sponge and water. If the article is to be exposed to the weather or much wear, the gilding may be varnished with copal varnish.

Burnish Gilding.—As previously stated, this process requires a specially prepared foundation upon which to lay the gold, and as the preparation of this foundation is a distinct trade, the furniture dealer or cabinetmaker seldom finds it necessary to undertake it, the articles coming to his hand ready-prepared for gilding; but as in repairing picture frames, cornices, mirror frames, etc., it frequently becomes necessary to renew the foundation, a comprehensive description of the whole process is given.

Preparing the Woodwork.—After smoothing and dusting the work, coat the frames in every part with boiling-hot parchment size, as previously described, then mix a sufficient quantity of whiting with size to the consistency of thick cream, and with it by means of a brush, coat every part of the frame several times, permitting each coat to become perfectly dry before proceeding with the next. The wood will thus be covered with a layer of hard whiting nearly or quite a sixteenth of an inch in thickness. The size must not be too thick, and when mixed with the whiting should not be so hot as the preliminary coat of size.

Polishing.—When the preparations are quite dry, clean and polish them. To do this, wet a small piece at a time, and, with a smooth, fine piece of cloth, dipped in water, rub the part till all the bumps and inequalities are removed; and for those parts where the fingers will not enter, as the mouldings, etc., wind the wet cloth round a piece of wood, and by this means make the surface all smooth and even alike.

Where there is carved work, etc., it will sometimes be necessary to bring the mouldings to their original sharpness by means of chisels, gouges, etc., as the preparation will be apt to fill up all the finer parts of the work, which must be thus restored. It is sometimes the practice, after polishing, to go over the work once with fine yellow or Roman ocher; but this is rarely necessary.

Applying the Size.—Select the proper gold size from the recipes previously given; add parchment size until it will just flow from the brush; make it quite hot, and apply it to the wood with a very soft brush, taking care not to make the first coat too thick; let it dry and give two or three successive coats, after the last brushing it with a stiff brush to remove any inequalities. The work is then ready for the gold.

Laying the Gold.—The manipulation of the gold leaf has been described under the heading Oil Gilding. In the paint now being described, size used (being water size, which, as previously explained, is permitted to become hard and dry after being applied) must be moistened to cause the gold leaf to adhere to it. For this purpose, with a long-haired camel-hair pencil, dipped in water, go over as much of the work as you intend the piece of gold to cover; then lay the gold upon it in the manner previously explained. Be sure that the part to which the gold is applied is sufficiently wet; indeed, it must be floating, or the gold will be apt to crack. Proceed in this manner a little at a time, and do not attempt to cover too much at once, until by experience you are going to handle the gold with freedom.

Burnishing.—When the work is covered with gold, set it by to dry; there is a particular state or degree of dryness, known only by experience, in which the moulding is in a fit state for burnishing; it will probably be ready to burnish in about eight or ten hours, but it will depend on the warmth of the room or state of the air.

When it is ready, those parts intended to be burnished must be dusted with a soft brush; then, wiping the burnisher with a piece of soft wash-leather (quite dry), begin to burnish about an inch or two in length at a time, taking care not to bear too hard, but with a gentle and quick motion, applying the tool until all the parts of the surface are equally bright.

Matting or Dead Gold.—Certain portions only of the work are burnished, according to the fancy, and the facility with which the burnishing tool can be applied; the remaining parts are now to be deprived of their metallic luster, to make a more effective contrast with the burnishing. The parts thus treated are said to be matted or dead gold. The process is as follows:

Grind some vermilion or yellow ocher very fine, and mix a very small portion either with the parchment size or with the white of an egg, and with a very soft brush lay it evenly on the parts to be dulled; if well done, it will add greatly to the beauty of the work. Previous to matting, the work must be well cleared of superfluous gold, by means of a soft brush.

Finishing.—In elaborate works it is frequently impossible to lay gold leaf into all the intricacies of an elaborate design, and the parts thus left bare must be finished by touching up with a small brush charged with shell gold, or gold powder, mixed with gum Arabic to the proper consistency. The following recipe describes the preparation of shell gold:

Shelled Gold.—Take any quantity of leaf gold and grind it with a small portion of honey, to a fine powder, add a little gum arabic and sugar candy, with a little water, and mix it well together; let it dry.

Silver Size.—Grind pipe clay fine with a little black lead and good soap, and add parchment size as directed for gold size.

Composition for Frame Ornaments.—The ornaments for gilded mirror frames, etc., are usually moulded from some plastic substance that is somewhat tougher and more durable than the ordinary gilding foundation of whiting and size. The proper moulds being prepared, they are thoroughly rubbed upon the inside with sweet oil, and the composition firmly pressed in; after removing the mould the cast may be dried by a gentle heat, or while still plastic it can be applied in its proper place and bent into any position. Following are recipes for composition:

Dissolve 1 pound of glue in 1 gallon of water. In another kettle boil together 2 pounds of resin, 1 gill of Venice turpentine, and 1 pint of linseed oil; mix all together in one kettle, and boil and stir till the water has evaporated. Turn the whole into a tub of finely rolled whiting, and work till it is the consistency of dough.

Boil 7 pounds of best glue in 7 half pints of water. Melt 3 pounds of white resin in 3 pints of raw linseed oil. When the above has been well boiled put them into a large vessel and simmer them for half an hour, stirring the mixture and taking care that it does not boil over. The whole must then be turned into a box of whiting rolled and sifted, and mixed till it is of the consistency of dough.

To Manipulate Gold Leaf.—Get a piece of paper, thin enough to show shadow of gold leaf through, slightly wax it, lay it on gold leaf; the latter will then adhere, and can be easily worked and will come off clean. The paper should be slightly larger than the gold leaf, and the fingers passed over the paper to make the gold leaf adhere.

Bronzing.—This is a process for imitating on metal, plaster, wood, or other material, the peculiar appearance produced by chemical action upon the surface of bronze metal. It is accomplished by spreading over the surface of the material to be ornamented a very thin coating of bronze powder, which is caused to adhere either by applying it directly upon a coating of any of the sizes mentioned in the foregoing pages, or by mixing with a vehicle, such as gum arabic or transparent varnish. The latter is most desirable, as in the other case, being subject to the direct action of the atmosphere, the bronze powder soon tarnishes. In ornamenting furniture, bronzing is generally employed to represent gilding, a variety of bronze called gold bronze being used, which affords an excellent imitation, but is not very lasting. It is usually applied after the completion of the other finishing processes, the ground work being prepared in the manner described under Oil Gilding, and the size likewise applied as there described. A small wad of cotton batting is then dipped in the bronze and passed gently over the sized portions, causing the bronze to adhere. In the other method—that of applying the bronze by means of a vehicle—the preliminaries of whiting the ground and sizing are not necessary, a small quantity of bronze being simply mixed with the vehicle employed to such a degree of fluidity that it will flow easily, and in that condition applied with a fine brush. Many preparations are used as vehicles, such as transparent varnish thinned with turpentine, gum arabic dissolved in water, and gold size reduced with parchment size. There are a variety of colors in bronze powders, and to produce the best effect the size or vehicle should be of a color similar to that of the bronze used; in gold size the coloring pigment is ocher, and in its place, for green bronze, or blue bronze, may be employed respectively verditer, vermilion or Prussian blue, a very small quantity being sufficient. In bronzing on painted work the ground should be as nearly as possible the color of the bronze to be applied.

Banana Solution.—The so-called "banana solution" (the name being derived from its odor) which is used in applying bronzes of various kinds is usually a mixture of equal parts of amyl acetate, acetone and benzine, with just enough pyroxylin dissolved therein to give it sufficient body. Powdered bronze is put into a bottle containing this mixture and the paint so formed applied with a brush to the article to be bronzed. The thin covering of pyroxylin that is left after the evaporation of the liquid protects the bronze from the air and keeps it from being wiped off by the cleanly housemaid. Tarnished picture frames and tarnished chandeliers to which a gold bronze has been applied from such a solution will look fresh and new for a long time. Copper bronze as well as gold bronze and the various colored bronze powders can be used in the "banana solution" for making very pretty advertising signs for use in the drug store. Lettering and bordering work upon the signs can be done with it. Several stiff, very small painter's brushes are needed for such work and they must be either kept in the solution when not in use, or, better still; washed in benzine or acetone immediately after use and put away for future service. It is needless to add that as the "banana solution" is volatile it must be kept well corked.

STAINING WOODWORK WITH ACIDS

For staining wood brown, sulphuric acid, more or less diluted, according to the depth of stain desired, is applied to the wood, previously cleaned and dried with a brush, and when the acid has acted enough its further action is arrested by the application of liquid ammonia.

To age oak artificially, liquid ammonia is laid on with a rag or brush, which does the work rapidly and effectually.

To darken cherry, rub it over with nitric acid of 1.2 specific gravity, and after permitting it to stand for twelve hours, wash and dry thoroughly. Nitric acid gives a permanent yellow stain, which may be converted into dark brown by subsequent application of tincture of iodine.

A hot, concentrated solution of picric acid gives a very fine yellow effect. Aqua fortis, diluted with three times its own weight of rain-water, brushed over the wood, gives a more true yellow effect than the undiluted nitric acid (aqua fortis).

A bright golden yellow stain is made by digesting ¼ ounce of powdered madder for twelve hours in a ounces of sulphuric acid and then filtering through cloth. The articles to be stained should be immersed in the fluid for three or four days.

ON HARDWOOD FLOORS

The finish and care of hardwood or parquette floors has been and is now a source of great trouble and annoyance to housekeepers. In many cases where beautiful floors have been laid, they have been left to be finished by persons who have not troubled themselves with finding out the best method of finishing, and the usual way for such persons to do is to treat them with shellac or varnish, says a writer in one of our exchanges. This is all wrong, as a moment's thought will convince any one that a surface that is constantly walked over needs something different from the coating of gum that is left on the surface after the spirit used in dissolving the shellac or varnish is evaporated. This coating then becomes brittle, and is ground up into minute particles by the nails in the boots and swept away, leaving the wood bare, right where it is most exposed to view.

As a matter of course, the beauty of the floor is soon gone, and instead of being an attractive part of the furnishing, the sanitary consideration very often is about all that keeps one from nailing a carpet over the whole floor. Others use linseed oil, and everybody knows that an oil finish is one of the best methods of finishing wood, but the objection is, that each time the oil is applied it darkens the wood, and in a short time the different kinds of wood are of the same color.

Now the question arises, which is the true and only way of finishing floors properly? And the answer is, by the use of hard wax, which, however, must be so prepared that the trouble of applying it and the stickiness attending ordinary beeswax and turpentine are entirely obviated. The wax is treated with special liquids and made into a preparation.

Among the many different things tried, hard wax was found to be the most satisfactory in its results. It is so simple, that when once the floor has been properly filled and finished with it, any servant can renew and keep the floors fresh and bright as long as the wood lasts, and as it does not materially change the color, the wood always retains its beauty. An application about once a year is all that is necessary, if the floors are rubbed over, when a little dull, with a weighted brush or cloth.

In repolishing old floors that have been in use for a length of time and become dull looking, it is only necessary, after they have been cleaned, to rub on a thin coat of the hard wax finish with the brush or cloth, as stated above. If the floors have been varnished and the varnish is worn off in places, as mentioned above, the best way is to have the varnish scraped off, and then a thin coat of the hard wax should be applied and treated as the new wood after it is filled. But if it is inconvenient to have the floor scraped, or the expense too much, the main object being to restore the color in those places which are worn and defaced, the following mixture is recommended: One part linseed oil, 1 part liquid dryer and 2 parts turpentine; a cloth should be dampened with this and applied to the worn and defaced places, which will have the desired effect. After being wiped off clean, it ought to dry twenty-four hours, and then be polished with the hard wax finish. It is very important never to use the wax over oil that is not thoroughly dry, as the floor would invariably be sticky.

Finally, it would be well to mention that hardwood or parquette floors should never be washed with soap and water, as it raises the grain and discolors the wood. After the floors have been properly filled and finished with the hard wax, dirt will not get into the pores, but stays on the surface and consequently can be removed with a brush or cloth, or, if necessary, dampen cloth with a little turpentine. This will take off any stain from the finish.

An excellent method of waxing floors is as follows: Take 1 pound of the best beeswax, cut it up into very small pieces and let it thoroughly dissolve in 3 pints of turpentine, stirring occasionally, if necessary. The mixture should be only a trifle thicker than the clear turpentine. Apply with a rag to the surface of the floor, which should be perfectly clean. This is the difficult part of the work; for if too much or too little is put on a good polish is impossible. The right amount varies, less being required for a hard, close-grained wood, and more if the wood is soft and open-grained. Try a square foot or two at first. Put on what you think will be enough, and leave the place untouched and unstepped on for twenty-four hours, or longer, if needful. When thoroughly dry, rub with a hand brush. If it polishes well, repeat the whole process over the entire floor. If it does not, remove the wax with fine sandpaper, and lay again, using more or less than before, as may be necessary, and continue experimenting until the desired result is secured. If the mixture is slow in drying, add one part japan to six of turpentine.

Birch makes an entirely satisfactory floor for dancing.

as well as for kindred uses. It is easily brought to a smooth surface and a fine polish, and is of a rich amber color of an even shade. In addition, it has that rare elasticity and resiliency that make it alike delightful for walking and dancing. It costs about 10 cents laid, and is in no way a disappointment to those using it.

What is said of birch applies equally well to hard maple, both the white and the red varieties, the white being that chosen for floors, and is the lightest-colored of the woods so used. It is very hard, takes readily a fine polish; the boards are not liable to warp, but, unfortunately, require the very closest care in the drying to prevent shrinkage when laid. It is lasting, and is but little affected by water. Only beech, hickory and white oak approach it in lightness of color. Hickory has sterling qualities, too generally appreciated to need detailed discussion of its intrinsic worth, yet it is sadly neglected when the question of flooring is under consideration. Perhaps this is due to the difficulty with which it is laid. It is an open-grain wood, but takes polish with ease.

Beech makes almost an ideal floor, light-colored and hard, and has the rare quality of wearing smoother with age; at times it is found beautifully bird's-eyed. In the Southern States it grows in the greatest profusion in the swamps and lower woods, but is unappreciated, only enough being preserved for use in making plane stocks and other tools requiring a hard, durable wood that does not shrink, warp or split. It could be laid for 20 cents per foot. And along with it goes apple, which polishes to a rich, delicate amber color; the cost being about the same as beech, but the apple wood has the tremendous disadvantage of not being obtainable in large boards.

The laying of a hardwood floor requires not only a good carpenter, but an expert judge of woods, and of the individual boards, because only by carefully selecting and placing like planks can we get a permanently even surface. Suppose a plank of heart and one of sap should be placed side by side; no matter what the wood, when a rainy season may come the sap will swell more and rise above the heart. Even when they come from like relative positions in two like trees, their texture may differ so widely as to make them undesirable companions. In spite of the nicest workmanship and the best judgment in selecting, some inequalities of surface will be present till removed by the most thorough sandpapering. This should be done with enough care to avoid scratching; then comes the polishing.

To get the best results, hardwood floors should be laid after the building is thoroughly dry, and in case of new building it should be the last work done.

Care should be taken that the surface on which the floor is laid is clean and smooth. Drive the flooring well up and be careful not to break the tongue.

Seven-eighths-inch flooring should be nailed with 2½-inch special flooring nail. For ⅜-inch flooring 1¼-inch finishing nail, No. 15, will be found about right.

An oak floor after being laid should be evenly cleaned off and sandpapered until perfectly smooth. It must then be filled with what is known as "wood filler," and allowed to stand for six to ten hours. This filler can be made any shade desired.

If a wax finish is desired, apply two light coats of white floor shellac. Let the first coat stand one hour before putting on the second. After the second coat

has stood for two hours, sandpaper with No. o sandpaper and the floor is ready for the wax, an article
made expressly for this purpose and ready for use.

Put the wax on as thin as possible and let it stand
for half an hour, then with a rubbing brush rub across
the grain of the wood and again lengthways until the

FIG. 117

brush slips easily, then take a piece of soft carpet and
rub until the desired polish is obtained.

For maple, birch, or other close-grain woods, use
the same process, omitting the "wood filler."

Estimate of Material Required.—For laying and finishing ⅜ flooring per 100 feet surface: 100 feet ⅜ flooring, 2½ pounds finishing brads No. 15, 3½ pounds
wood filler (for oak only), 3 pints shellac, ½ pound

floor wax. For ⅞ flooring, 6 pounds 2½-inch flooring nails will be sufficient per 100 feet.

A weighted brush with a long handle is generally employed for polishing a wax-finished floor, similar to the one shown in Fig. 117. The wax is applied with a rag or brush, after the filler has been properly rubbed down and all is hard and dry. The weighted brush is then rubbed over the surface to and fro until the desired polish is attained.

Stained Floors.—A floor stained to represent dark old oak is preferred by many. The mixture for accomplishing this is sold at all paint shops, and comes in grades 1, 2, 3, and 4, varying from light to dark. If the boards are smooth and fine-grained, a satin wood or pitch pine stain or polish is preferred; but if the floor is old or rough it is folly to attempt any stain except that of dark oak or dark mahogany. Some of the mixtures used for this can be put on with a rag, although a brush is better. Pour the liquid into a saucer, dip the brush in, saturate thoroughly, rub evenly over the wood, and dry instantly with a soft cloth.

For the ultra-fashionable floor, which is of a pale shade of oak, sized and varnished, buy the desired amount of raw sienna powder; mix with water, and rub into the boards as directed above. Mahogany staining: Make a mixture containing ½ pound of madder, 2 ounces of logwood chips, boiled in 1 gallon of water; brush this over the wood while hot. When dry, go over this with a solution of pearlash, 2 drachms to 1 of water, size and polish. If a redder shade is required, it can be produced by smearing the surface with a strong solution of permanganate of potash, which is left on for five minutes. The wood is then carefully washed, dried and polished.

A good cheap oak stain is made of equal parts of potash and pearlash, 2 ounces of each to a quart of water. As potash is a solvent, care must be taken to keep it from the hands; and an old brush should be used.

For other stains and methods of applying them, see recipes described in previous pages, where stains for nearly all purposes are given.

MISCELLANEOUS MATTERS

Floor Polish.—Cut beeswax into small pieces, or else grate it up; add turpentine, and allow the mixture to stand for twelve hours; then heat it over the fire till it dissolves. Care must be taken not to heat the mixture too hot, and also the flame must not come too near, for explosive vapors are generated, which are liable to catch fire.

Dull Polish on Stained Whitewood.—The dull polish that is seen on most furniture is obtained by partly French polishing the article, and then removing any apparent shine or gloss by well brushing the surface over with medium grade pumice-powder or fine emery; or the stained wood might be coated with spirit varnish. In the absence of details as to the purpose for which the stained wood is to be used, no other procedure can be suggested. Stained floor-boards, for instance, would not require French polishing, nor even spirit varnishing, because a suitable polish can be readily obtained by using beeswax dissolved in turpentine, applied with a weighted brush. On the other hand, on furniture goods French polish serves a double purpose; the polish partly fills the grain or pores of the wood, and gives a hard surface that can be dulled without rubbing off the stain.

Refinishing Oak Doors that are Badly Weather-Stained.—If possible, take the doors off the hinges and lay them down flat on some trusses or boxes, and remove the old varnish with ammonia or a mixture of 2 parts strong ammonia and 1 part of turpentine and benzine, using a stubby brush to get into the cutwork and about the mouldings. When all the varnish has been removed, dope over stained portions with a strong oxalic acid solution, and see whether you cannot bleach the wood by that operation. If this will not work, you have to resort to staining. Use raw sienna for light effect, and, after staining, use paste wood filler, colored to match the stain. Then proceed as you would on new work. If the light stain does not hide the weather stains, you will be obliged to use a darker stain and darker filler.

Coloring Wood Clear Through.—All the sap is expelled and the log is then treated with chemicals, and the color or colors are pressed into the wood. Any shade desired can be obtained, and, in fact, several colors can be merged one into the other, producing a very beautiful effect. On cutting up the samples we received, we found that the color was evenly distributed all through the fibers, the grain of the wood giving a very pleasing effect, especially when polished. The wood, it is claimed, dries sooner than by ordinary seasoning, and it can also be rendered fireproof by adding special chemicals. Of course, painting is done away with, so that the natural structure of the wood is seen to better advantage than when painted in the ordinary way. The coloring is, we understand, free from arsenic and quite harmless; the colors do not fade, and, of course, cannot be worn off by rubbing, etc.

Cleaning Polished Wood.—A good encaustic, which will clean and polish at the same time, may be made from wax, sal soda and any good soap. The wax and soap should be shaved and dissolved in boiling water. Stir frequently and add the soda. Put the mixture in something which may be closely covered, and stir constantly until cool. This may be applied to floors, furniture, marbles, tiles, bricks, etc. It will remove ink from polished surfaces. The French use white wax on white marbles, but this is not absolutely necessary.

Finishing Hardwood.—If it is open-grained wood I should first fill it with paste filler, then I would give it a coat of shellac, and after that I would bring it up with a first-class varnish.

It would be all right to finish it all in shellac if it could be kept from moisture, but wherever a drop of water touches a shellac finish it will turn white. And just as like as not the mistress will set the servants to wiping up the hardwood finish with a damp cloth. Now a good varnish will stand it, but shellac won't. But the best way to clean furniture and hardwood work is to use crude oil—only a very little of it—and then wipe it off thoroughly with cotton waste or cheesecloth. The latter is preferable because it has no lint to catch on the woodwork, although if you rub it dry enough with cotton waste you can rub off any lint that may be left. The crude oil acts as a varnish renewer as well as a cleaner. But if it is not thoroughly wiped off with plenty of elbow grease it will catch the dirt and look pretty bad. Crude oil is a good thing, provided you don't use too much of it, and then, again, provided you don't leave it on.

Making Paste Wood Fillers.—Paste fillers for hard woods are made from any of the following materials, or a combination of these: silex or silica, terra alba, whiting, china clay, starch, rye flour, and sometimes barytes. Silex or terra alba will, on drying, give the least discoloration to the wood. The pigment should be of impalpable fineness and intimately mixed to a stiff paste with one-third each of pale linseed oil, pale gold size japan and turpentine. This paste may be either run through a mill or be given a very thorough mixing, and to test it for quality it should be thinned with turpentine to the consistency of a varnish, applied with a varnish brush to open-grained wood, preferably oak, allowed to set for about twenty to thirty minutes, and the surplus filler removed by wiping across the grain in the usual manner. After twenty-four to thirty-six hours, the surface should be lightly sand-papered and a good, flowing coat of rubbing varnish applied, which, when fairly well set, should not show any pitting or pin holes. Should it pit, however, or show pin holes or needle points, the filler is defect-ive in binding properties, and the portion of japan should be increased, with a corresponding decrease in the proportion of turpentine. The linseed oil and the gold size japan must be of good body, and if corn-starch or rye flour is used in connection with silex or silica, the proportions should be about one of the former to five of the latter by weight.

Filler for White Ash.—As white ash is a very porous wood, it should be treated with an extra light mineral paste wood filler, made from clean silex, mixed with 2 parts bleached linseed oil, 3 parts pale japan gold size and 1 part turps, to stiff paste and thinned for use with turpentine to the consistency of medium-bodied varnish. When dry and hard the surface should be smooth sandpapered and given a coat of white shellac varnish, after which it may be finished with rubbing varnish, that may be rubbed and polished in the ordinary way.

Good Wood Finish.—Richness of effect may be gained in decorative woodwork by using woods of different tone, such as amaranth and amboyna, by inlaying and veneering. The Hungarian ash and French walnut afford excellent veneers, especially the burs or gnarls. In varnishing, the varnishes used can be toned down to match the wood, or be made to darken it, by the addition of coloring matters. The patented preparations, known as "wood fillers," are prepared in different colors for the purpose of preparing the surface of wood previous to the varnishing. They fill up the pores of the wood, rendering the surface hard and smooth. For polishing mahogany, walnut, etc., the following is recommended: Dissolve beeswax by heat in spirits of turpentine until the mixture becomes viscid; then apply, by a clean cloth, and rub thoroughly with a flannel or cloth. A common mode of polishing mahogany is by rubbing it first with linseed oil, and then with a cloth dipped in very fine brickdust; a good gloss may also be produced by rubbing with linseed oil, and then holding trimmings or shavings of the same material against the work in the lathe. Glass-paper, followed by rubbing, also gives a good luster.

There are various means of toning or darkening woods for decorative effect, such as logwood, lime, brown soft soap, dyed oil, sulphate of iron, nitrate of silver exposed to sun's rays, carbonate of soda, bichromate and permanganate of potash, and other alkaline preparations are all used for darkening woods. The last three are specially recommended. The solution is applied by dissolving 1 ounce of the aikali in 2 gills of boiling water, diluted to the required tone. The surface is saturated with a sponge or flannel, and immediately dried with soft rags. The carbonate is used for dark woods. Oil tinged with rose madder may be applied to hardwoods like birch, and a red oil is prepared from soaked alkanet root in linseed oil. The grain of yellow pine can be brought out by two or three coats of japan much diluted with turpentine, and afterwards oiled and rubbed. To give mahogany the appearance of age, lime water used before oiling is a good plan. In staining wood, the best and most transparent effect is obtained by repeated light coats of the same. For oak stain a strong solution of oxalic acid is employed; for mahogany, dilute nitrous acid. A primary coat, or a coat of wood fillers, is advantageous. For mahogany stains the following are given: 2 ounces of dragons' blood dissolved in 1 quart of rectified spirits of wine, well shaken; or raw sienna in beer, with burnt sienna to give the required tone; for darker stains boil ½ pound of madder and 2 ounces of logwood chips in 1 gallon of water, and brush the decoction while hot over the wood. When dry, paint with a solution of 2 ounces of potash in 1 quart of water. A solution of permanganate of potash forms a rapid and excellent brown stain.

Easy Method of Finishing Woodwork.—French polishing as a means of finishing furniture and woodwork is generally regarded as a most tedious operation, owing to the number of solutions to be used on work that is built up of various kinds of wood, in bringing it up to uniform color, and in polishing it so as to bring out and reflect to the fullest extent the markings or figure of the wood. On high-grade goods, with a bright, lustrous, level finish this is so. Yet much furniture is not of high-grade finish, so far as the polisher is concerned; for instance, bedroom furniture that is stained green is rarely finished out extra bright, and the same may be said of fumed oak goods and many American organs. In fact, some goods look far better with a faintly lustrous polished surface than if finished out very bright, especially if the surface is at all uneven or badly cleaned up. A process of finishing known as "dry shining" strikes a medium between high-grade finish and simple spirit varnishing. In the crudest form of this process the work is simply oiled and a wet rubber of polish applied all over, not sufficient being used to fill the grain, but just enough to kill the oil. This treatment is generally considered good enough for the insides of drawers, cupboards, etc., the object being to remove an unfinished appearance and to prevent the surface getting as dirty as it otherwise might. From this better degrees of finish may be reached. The work may be oiled, filled in, one or more rubbers of polish laid on just to fill up the grain, and then an even coat of spirit varnish applied. If the articles are of white wood, they may be stained to imitate some choicer wood before oiling; and if the goods are likely to be subject to hard wear, the coating of spirit varnish may be omitted, the polish being worked out fairly dry to ensure the removal of all oil; then apply a coat of oak or painter's varnish, which, however, gives a bright surface when dry, and is merely mentioned as a means of obtaining a bright finish with the minimum of trouble.

Egg-shell finish also does not require the trouble

some operation of spiriting out. Here the work is brought up to a stage nearly approaching that for spiriting, but the surface of polish when hard is dulled by rubbing or brushing with fine-grade pumice-stone powder or flour emery, in which condition it may be left. If a gloss instead of a shine is preferred, the wood should have a smart rubbing of beeswax and turps. Black work has a specially chaste appearance thus finished, and the black stain of logwood and iron solution may be used, aniline spirit black being employed for imparting density of color to pale shellac polish. If it is not convenient to use varnish, and a simple solution of shellac in spirits (4 ounces orange shellac dissolved in 1 pint methylated spirit) is the only solution at hand, a passable finish may still be gained by enclosing the pad in a piece of soft rag and finishing out by working it in straight lines, after a body has been put on without a covering. When the articles must be stained, it will be found more economical to buy the stains ready-made if only a small quantity is required. Dry shining has at least the merit of building up a surface that can be taken in hand again and French polished.

Metallization of Wood.—Some artisans in Germany have succeeded in turning to practical account the recently devised process by which wood is made to take on some of the special characteristics of metal, that is, the surface becomes so hard and smooth as to be susceptible of a high polish, and may be treated with a burnisher of either glass or porcelain; the appearance of the wood being then in every respect that of polished metal, having, in fact, the semblance of a metallic mirror, but with this peculiar and advantageous difference, namely, that, unlike metal, it is unaffected by moisture. To reach this result the wood is steeped in a bath of caustic alkali for two or three days, according to its degree of permeability, at a temperature of between 164° and 197° Fahr.; it is then placed in a second bath of hydrosulphate ot calcium, to which a concentrated solution of sulphur is added, after some twenty-four or thirty-six hours; the third bath is one of acetate of lead, at a temperature of from 95° to 122°, and in this latter the wood is allowed to remain from thirty to fifty hours. After being subjected to a thorough drying it is in a condition for being polished with lead, tin or zinc, as may be desired, finishing the process with a burnisher, when the wood apparently becomes a piece of shining, polished metal.

How to Tone Down New Mahogany, Oak, etc.—In making repairs to furniture, it usually happens that the new wood is considerably lighter in tone than the old, and ordinary stains will not match it so as to give satisfaction. This can be done easily, however, by means of a solution of bichromate of potash. To make this, purchase a cent's worth of the chemical, and placing it in an ordinary medicine bottle, fill up with water and shake until dissolved. To use the solution, rub a small quantity on the wood to be darkened, and await results. If not dark enough, give another coat. It dries in a few minutes, and can be sandpapered after, as it is not a surface stain, but a chemical one. By a judicious use of the above solution it is easy to match old work of any description, so that the new and old cannot be distinguished from each other.

Spirit Varnish for Violins.—Spirit varnish is difficult to apply evenly, owing to its drying so quickly. The color generally appears streaky. In any case, no shellac should be used in the varnish, as shellac is too hard. A good spirit varnish is made as follows: First size the violin with a mixture of 3 parts of best copal varnish and 1 part of turpentine, applied hot with a rag, and well rubbed in. Color ½ pint of alcohol with turmeric and a little red sanders added to take away the greenish tinge. Dissolve 2 ounces of gum sandarach (juniper) in ½ pint of alcohol. Put the two half-pint mixtures together, and add 2 tablespoonfuls of Venice turpentine and 2 ounces of white shellac. When dissolved, filter through cotton wool.

Putting Transfers on Coach Panels.—The method of transferring crests and monograms to the panels of coaches, etc., is as follows: Cover carefully the face of the design (that is, the colored or printed side) with a thin, smooth coat of gold size mixed with two or three drops of varnish, being careful to cover all parts that are to be transferred. Let the gold size coating become thoroughly tacky or sticky, then lay the design face downward on the panel to be decorated, and roll it down smoothly with a rubber roller, pressing out all air bubbles. When the adhesive has got quite dry, thoroughly soak the paper with water by means of a sponge, then gently peel off the paper from one corner. Sponge the surface composition off the panel, and when the design is quite dry, apply a finishing coat of varnish. Transfers can also be applied without coating them with gold size, if the panels have been recently varnished, and have a good tack (that is to say, when the varnish is sticky , as the design will then adhere by gentle pressure.

Paint for Blackboards.—The best blackboard paint is made by moistening 4 ounces dry lampblack with alcohol, rubbing it out with a spatula, gradually adding 1 quart of shellac varnish, and stirring into this 3 ounces flour of pumice and 3 ounces finely pulverized rotten stone; then straining through a fine sieve or strainer to break up any lumps that may have formed. This is applied quickly to the bare wood, so that no laps are formed, and in a day or so a second coat may be applied, and after standing a day or two longer may be haired or mossed.

Ebonizing.—Apply to the wood, by means of a brush or sponge, a solution of hydrochloric aniline dissolved in water, to which has been added a little protochloride of copper. When this coating has dried, apply similarly a solution of bichromate of potash dissolved in water. After this process has been repeated two, or at the most three, times, the wood will assume a clear, full, durable black color, which is affected neither by the action of light nor dampness.

Polishing Boxwood Draughtmen.—The cheaper class of draughtmen are simply coated with a good quality spirit varnish, but high-grade goods are polished in the lathe. The polish that is used and the method of applying the polish differ slightly from the method that is employed in polishing flat surfaces. A bright finish on both sides and edges is only obtained after several handlings, the chief difficulty being the manipulation in the early stages, such as the provision of suitable chucks, the avoidance of the use of glass-paper, and the knack of using the polish so that it will not clog up the finer grooves. If ordinary French polish is used, it should not be applied with new wadding; a wad made from a rubber that has been used on other work should be employed, so that there may be less risk of loose fluff sticking to the work while the polishing is being done. The wad would not require the rag covering that is usual on flat surfaces. If a lathe is not available, very good results could be obtained by using polish for sealing up the pores of the wood and forming a smooth foundation, and then applying carefully a coating of good quality clear spirit varnish. Black goods should be stained first with French black water stain, and the polishing done with black polish. White polish made from bleached shellac, or a transparent polish, should be used in preference to polish that is made from orange or lemon shellac.

Softening Putty.—To soften putty that has become hard by exposure, so as to remove it easily from a sash, take 1 pound of pearlash and 3 pounds of quickstone lime; slake the lime in water, then add the pearlash, and make the whole of about the consistency of paint; apply it to both sides of the glass, and let it remain for twelve hours, when the putty will be so softened that the glass may be taken out of the frame with the greatest facility.

Bruises in Wood.—To take out bruises in furniture, wet the part with warm water; double a piece of brown paper five or six times, soak it, and lay it on the place; apply on that a hot flatiron till the moisture is evaporated. If the bruise be not gone, repeat the process. After two or three applications, the dent or bruise will be raised level with the surface. If the bruise be small, merely soak it with warm water, and apply a red-hot poker very near the surface; keep it continually wet, and in a few moments the bruise will disappear.

Wood Stains.—The following have been published by a German paper as formulæ for some wood stains, which may be put up in a dry form, and when wanted for use may be readily dissolved in water: Oak wood; 5 kg. of Cassel brown, .5 kg. of potash, and 10 kg. of rain-water, boiled together for an hour, the whole strained through a linen cloth, and the clear, dark-colored liquid boiled to a syrupy consistency. Walnut wood: A decoction of Cassel brown, 3 kg.; potash, .3 kg.; and water, 7 kg.; the whole strained through linen, and during evaporation to syrup 2.5 kg. of extract of logwood added. Mahogany: A decoction of extract of Brazil wood, 3 kg.; potash, .25 kg., and water, 3 kg.; to which, before evaporating to syrup, 150 gr. of eosine are added. Ebony: 5 kg. of extract of logwood, boiled with 11 kg. of water, and, when near the syrupy state, 300 gr. of iron nitrate added; evaporated to a syrup under constant stirring. All the above stains are brought into a dry condition by running the respective syrups into trays of sheet iron, with low rims, in which the syrup hardens, and is afterward broken up and ground.

It is often desirable to retain the grain of the natural wood exposed to view, at the same time to preserve its surface from decay and give it a more beautiful appearance; this is done either by polishing or varnishing. To varnish such woods a little skill is required to obtain a really good gloss, smooth as glass, upon its surface. All roughness should be carefully removed, being particular not to leave any marks, especially across the grain, of the sandpaper or other material used in smoothing, and the work should be afterwards well sized, either with gelatine or good glue size. This size is to prevent the absorption of the varnish in soft places, and to obtain a more even gloss. Sizing sometimes has a tendency to raise the grain of the wood, more particularly of soft wood, especially if applied warm. Use oak varnish.

Aniline Dyes.—Aniline dyes are of two kinds, one dissolving in water, the other in spirits. As they have a tendency to fade in the light, the water dyes are preferable, as they can be mixed with a little vinegar, this greatly hindering the fading out process. To dissolve in spirits, use a spirit varnish, such as painters use. No definite amount necessary to stain varnish can be given, and it will be necessary to experiment with it.

HARDWOOD FINISHER.

QUESTIONS.

1. Give a description of the different kinds of wood that used to be in vogue about thirty or forty years ago, but are now reckoned inferior in the manufacture of furniture.

2. Whether have dark or light colored woods the preference from an æsthetic point of view?

3. Mention some of the valuable qualities to be found in white oak.

4. Mention other two kinds of wood that are frequently used for high-class work, and well adapted for finishing purposes.

5. Whether is it preferable a finish in hardwood or a finish in pine?

6. Is there any difference in cost between a finish in the best clear pine and the best selected hardwood?

7. What is essential in the choice of all kinds of hardwood for finishing purposes?

8. What are the characteristic features in hardwood that recommends it above others?

9. What class of work is pine peculiarly adapted for?

10. What other kinds of soft wood are fairly good for finishing purposes?

11. Mention the names of some of the woods that have all coarse grain, and that are not so suitable for tasteful work.

12. What has been the result of introducing the modern methods of polishing finished woodwork?

13. Give a description of the process termed "French polishing" and for what purpose it is best adapted.

14. Give an account of the organic tissue in woods, also the variable organic elements associated with it, and examples illustrating same.

15. Give a description of the exterior characteristics of woods, and their subdivision into two classes, also the names of some of the woods in each class.

16. Give a description of the different ways in the consideration of the density of wood.

17. Whether is the density of the harder or softer woods more preferable and popular with wood finishers, and state the reasons for preference?

18. Give a description of the qualities of "walnut" wood.

19. Give a description of the qualities of the "mahogany" wood, and the several kinds.

20. Give a description of the qualities of the "cherry" wood.

21. Give a description of the qualities of the "black birch."

22. Give a description of the characteristic features to be found in the different varieties of oak.

23. Give a description of some of the qualities to be found in the "butternut."

24. Give a description of the qualities to be found in "rosewood."

25. Give some of the characteristics that are to be found in the "apple" wood.

26. Give a description of some of the qualities to be found in the "maple" wood.

27. Give a few of the characteristic qualities to be found in the "chestnut" and "ash."

28. When does it seem superfluous to have complex decoration in the finishing of hardwoods?

29. What wood is made most use of for interior finish, and what are some of the articles for which it is well adapted?

30. Describe the difference between the working of oak, particularly in the framing up of panel work, from ordinary pine or other soft wood panel work.

31. Give a description which shows the method of setting out the twist or spiral for a column, pillar or spindle.

32. Give a description of the method of making properly a dovetail joint.

33. Give a description of secret lap dovetailing, and for what purposes it is well adapted.

34. What is the difference between secret and plain lap dovetailing?

35. Give a description of what is mean by the process "miter dovetailing."

36. Give a description of what is meant by the process "bevel or splay dovetailing."

37. Give a description of the method adopted in the manufacture of veneered doors, when a number of them are to be made at one time.

38. What should be done in the manufacture of first class doors before they are veneered?

39. What equipment is required aside from the usual door-making machinery?

40. Give a description of the preparatory work of the materials previous to their construction.

41. Give a description of the process of construction in the manufacture of doors and particulars regarding the veneers and process of veneering.

42. Give a description of the advantages derivable in the construction of "dowel" doors.

43. Give a description of the best method whereby to test the quality of glue.

44. Give a description of the best way in which to prepare glue for use.

45. Give a description of the process in applying the glue to the purposes for which it is intended.

46. Give a description of the kinds of wood suitable for veneering purposes, and the preparatory processes necessary.

47. Give a description of the process of "jointing" in veneering.

48. Give a description of the process of "veneering by caul."

49. Give a description of the process of "veneering round and tapering columns."

50. Give a description of veneering small work by

using the cauls, such as in making picture frames, clock stands or similar work.

51.   Give a description of what should be done preparatory to the process of polishing.

52.   Give a description of "the scraper," for what purposes it is employed, and the method of its manipulation.

53.   Give a description of the proper method of sharpening scrapers.

54.   Give a description of the method of using sandpaper.

55.   Give a description of "rasps and files" and their uses and modes of manipulation.

56.   Give a description of the different kinds of saws for working hardwood, and how to manipulate them.

57.   Give a description of the method of sharpening tenon saws.

58.   Give a description of planes in general, and the methods of manipulating them.

59.   Give a description of the grain direction for planes.

60.   Give a description of the proper method of setting an iron in a plane.

61.   Give a description of oilstones for sharpening plane irons, and method of manipulating the iron during the process of sharpening.

62.   Give a description of the process termed "secret or blind nailing."

63.   Give a description of what the term "wood fillers" means, and the methods of their application.

64.   Give a description of the preparatory work necessary before commencing the process of "filling in."

65. Give a description of the principal fillers used in the trade.

66. Give the description of a walnut filler for medium and cheap work.

67. Give the description of a walnut filler for imitation wax finish.

68. Give the description of a walnut filler for first class work.

69. Give the description of a filler for light woods.

70. Give the description of a filler for cherry wood.

71. Give the description of a filler for oak wood.

72. Give the description of a filler for rosewood.

73. Give a description of the operations to be employed when the work with the filler is done.

74. Give a description of the method in applying "luxeberry" to the wood.

75. Give a description of the process of wood staining in general, stating some of the varieties, and the best woods that are well adapted for their application.

76. Give the reason why French polishing is employed in finishing first class work.

77. Give a description as to the time and temperature in which the varnishing should be done, in order to be of durable character and produce beautiful work.

78. Give an example of how to treat cabinet work during the process of varnishing.

79. Give a description of how to manipulate the brush while varnishing, and the best kind of brush for the purpose.

80. What should always be considered before beginning the process of varnishing.

81. Give a description of the preparatory work necessary for the process of French polishing.

82. Give a description of the "pad or rubber" with which the polish is applied, and the best kind of material of which it should be made.

83. Give a description of the amount of polish to be applied to the "rubber," and the manner of manipulating the latter.

84. Give a description of what should be done with the old "rubber" when the job is finished upon which it was used.

85. Give a description of the time to be allowed for the polish to sink, and the process to be employed before commencing to polish again.

86. Give a description of the ingredients that compose a good all-round polish that can be relied on.

87. Give a description of how to use the material for bodying.

88. Give a description of what should be done when the rubber dries.

89. Give a description as to how long the first bodying-in process should be continued.

90. Give a description as to the number of times the work will require to be bodied.

91. Give a description of how to proceed before beginning to work a fresh body on a previous one.

92. Give a description of what should be observed by polishers when bodying up.

93. Give a description of the important matter regarding how to dry the rubbers.

94. Give a description of the final operation in French polishing, by which the gloss is put on the body previously applied.

95. Give a description of the process known as wax polishing.

96. Give a description of the class of wood upon which wax polishing is often applied, and the characteristic features it imparts in comparison with French polish.

97. Give a description of the appearance that wood stained black has after it is wax polished.

98. Whether upon coarsely grained woods or light woods of close texture, is wax polish better to be applied?

99. Give a description of the ingredients in the composition of wax polish, and the process of their admixture.

100. Give a description of the process of oil polishing, for what purposes it is best adapted, and the characteristic features in its favor.

101. Give a description of the process known as "dry shining," the method of its adaptation, and the chief advantages in connection with it.

102. Give a description of repolishing and reviving old work, and the various processes that should be adopted in the class of work to be operated upon, so that the best results may be produced.

103. Give a description of how to renovate the polish on German pianos, and the different ingredients and method of admixture that are applied to render the operation effective.

104. Give a description as to matching up satin wainut.

105. Give a description of how the wavy appearance of some woods may be given, and how veins, either black or red, may be produced in matching.

106. Give a description of the operation entailed in the final stage, when finishing off repolished work.

107. Give a description of the utility of dry colors, known as pigments, in the polisher's operations.

108. Give a description of the preparation of a mixture which is used in making an imitation marble which wears well, and its effective appearance when produced.

109. Give a description of the manner of finishing oak in general.

110. Give a description of how to produce the effect of a good imitation of antique oak.

111. Give another description of how a very clever imitation of the general antique can be obtained.

112. Give a description of how oak may be fumigated, stating the liquid used, and the method of procedure in the operation.

113. Give a description of how to darken oak.

114. Give a description of the different styles of oak finishes, namely, "bog oak," "weathered oak," "Antwerp oak," "black Flemish oak," "brown Flemish oak," "Malachite," and "Tyrolean oak."

115. Give a description of how to obtain a good "golden oak" finish.

116. Give a description of cherry wood and the best method of making it look like mahogany.

117. Give a description of black birch, and what woods it can be easily stained to resemble.

118. Give a description of mahogany, and the excellent qualities it possesses, also what may be done to darken the reddish hue which newly wrought mahogany presents.

119. Give a description of the process in repolishing and reviving old mahogany work.

120. Give a description of walnut and how it may be treated in the finishing, also mention some of the woods that may be stained to resemble it.

121. Give a description of the cypress wood, and its adaptability for the process of finishing, also the names of several of the varieties.

122. Give a description of rosewood and how it may be treated in the process of finishing.

123. Give a description of the maple wood, and its adaptability for staining purposes, stating some of the imitations that may be obtained.

124. Give a description of the maple wood, and its adaptability for finishing, also the method of obtaining an "egg-shell" gloss, "a dull finish," and "a polished finish."

125. Give a description of white and black ash, and its adaptability for finishing, also the method of obtaining an "egg-shell" gloss, "a dull finish," and a "polished finish."

126. Give a description of dyeing wood, and for what purpose this process is mostly applied.

127. Give a description of the process termed "gilding" and the characteristics of the design upon which it is employed.

128. Give a description of the process known as "burnish gilding."

129. Give a description of the process termed "bronzing" and how it may be accomplished.

130. Give a general description of the "metallization of wood," and the process by which it may be obtained.

HARDWOOD FINISHER

HARDWOOD FINISHER

INDEX TO PART TWO

INDEX TO PART TWO

INDEX TO PART TWO

www.ingramcontent.com/pod-product-compliance
Lightning Source LLC
Chambersburg PA
CBHW011301210326
41599CB00035B/7084